Frank Rattay

Electrical Nerve Stimulation

Theory, Experiments and Applications

Springer-Verlag Wien New York

Univ.-Doz. Dipl.-Ing. Dr. Frank Rattay
Technical University Vienna
Austria

Printed in Austria
Printed on acid-free paper

With 136 Figures

Cover-design: T. Erben, Wien

ISBN 3-211-82247-X Springer-Verlag Wien New York
ISBN 0-387-82247-X Springer-Verlag New York Wien

PREFACE

Functional electrical stimulation is the most important application in the field of clinical treatment with currents or magnetism. This technique generates artificially neural activity in order to overcome lost functions of paralyzed, incontinent, or sensory handicapped patients. Electricity and magnetism is also used in other cases, e.g., to stimulate bone growth or wound healing. In contrast to these applications, the basic mechanism of the artificial excitation of the nerve and the muscle fibers has become known in the last few years.

Although many textbooks are concerned with the natural excitation process, information on the influence of an applied electric or magnetic field is lacking. This book, written for students and biomedical engineers, should close this gap and, furthermore, it should stimulate the desire to design new instrumentation using optimal strategies.

That this book's English reads naturally, is due to the corrections of Kris Kingery and Robert Motz. I also wish to thank the many friends and students who have provided valuable hints, discussions and computations.

Vienna, September 1990 Frank Rattay

CONTENTS

ABBREVIATIONS AND SYMBOLS *

Ach	actetylcholine
AP	action potential
BVF	Bonhoeffer-Van der Pol-Fitzhugh (model)
C, C_m	capacity of cell membrane
c, c_m	capacity of cell membrane per cm^2
Ca^{++}	calcium ion
$[Cl]_i, [Cl]_o$	cloride concentration, inside and outside
CRRSS	Chiu-Ritchie-Rogart-Stagg-Sweeney (model)
d	fiber diameter
D	fiber diameter of myelinated axons, outer diameter including myeline sheet
E	voltage, (V is also used for voltage) **
f	activating function
FH	Frankenhaeuser-Huxley (model)
G_a	inneraxonal conductance
G_m	membrane conductance
g_{Na}, g_K, g_L	maximum conductance of sodium, potassium and leakage per cm^2 of membrane
h	probability for an ionic membrane gating process
HH	Hodgkin-Huxley (model)
I	current
I_{ionic}	ionic current
I_{el}	electrode current
I_{inj}	injected current
I_m	membrane current
I_{Na}, I_K, I_L, I_P	specific ionic currents (sodium, potassium, leakage, late sodium)

* Abbreviations including values for constants and parameters are listed in the next table.

As we use the common symbols it sometimes occurs that the same symbol has different meanings, but there is no problem with confusion, e.g., R stands for resistance but also R is a gas constant.

** Sometimes different symbols are used to be in agree with well known formulas found in literature.

i	current density, i.e., current per cm^2
i_{ionic}	ionic current density passing through the membrane
i_{Na}, i_K, i_L, i_P	sodium, potassium, leakage and late sodium current densities
i_{st}	stimulating current density
IH	interspike (time) histogram
$[Na]_i, [Na]_o$	sodium concentration, inside and outside of a fiber
$[K]_i, [K]_o$	potassium concentration, inside and outside
k	temperature coefficient
m	probability for an ionic membrane gating process
ms	millisecond
n	probability for an ionic membrane gating process
nm	nanometer $1nm=10^{-9}m$
p	probability for an ionic membrane gating process
PH	period (poststimulus) histogram
Q_{10}	acceleration factor for a temperature increase of 10°
R	resistance
R_e	resistance of extracellular medium
R_i	resistance of inside medium (axoplasm)
R_m	membrane resistance
r	distance
SE	Schwarz-Eikhof (model)
T	temperature
t	time
x	length coordinate of axons
Δx	segment length of fiber, for myelinated fibers Δx is equal to the internodal length
V	Volt; Voltage, often used for the reduced voltage across the membrane, i.e., in the resting state $V=0$
V_e	extracellular Potential
V_i	intracellular Potential
V_{Na}, V_K, V_L	Voltages across the membrane, caused by different ionic (sodium, potassium, leakage) concentrations at the inside and the outside of the fiber
V_{rest}	resting voltage across the membrane
z	distance of a small electrode to the centerline of an axon

$\alpha_m, \alpha_n, \alpha_h, \alpha_p$	coefficients in membrane models, e.g., in the HH or FH model
$\beta_m, \beta_n, \beta_h, \beta_p$	coefficients in membrane models, e.g., in the HH or FH model
β	time transformation factor
λ	space constant
$\mu m, \mu s$	micrometer, microsecond
τ	time constant

TABLE OF IMPORTANT CONSTANTS AND TYPICAL PARAMETERS *

c	capacity of cell membrane, $c \sim 1 - 2.5 \mu F/cm^2$
e	basis of natural logarithm
F	Faraday constant, $F = 9.64845 \cdot 10^4 C/mol$ the charge of one mol of single valenced ions.
L	nodal gap width, $L \sim 1 - 2.5 \mu m$
R	gas constant, R=$8.31441 J/(mol.K)$
r_m	resistance of resting cell membrane $r_m \sim 1 k\Omega$
ρ_i	resistivity of axoplasm $\rho_i \sim 50 - 200 \Omega \cdot cm$
ρ_e	resistivity of extracellular fluid $\rho_e \sim 300 \Omega \cdot cm$

* Data is also found as a concentrated form in the Tables and Boxes. — As biological data varies, its magnitude is symbolized by $\sim$.

1. FUNCTIONAL ELECTRICAL NERVE STIMULATION: A WAY TO RESTORE LOST FUNCTIONS

Introduction - historical remarks

Today electrical nerve and muscle stimulation has a broad field of application which includes the following:

- cardiac pacemakers
- phrenic pacemakers against respiratory insufficiency
- motor nerve stimulation for the paralyzed
- electrical stimulation for urinary or anal incontinence
- improvement of spasticity
- visual prostheses for the blind
- auditory prostheses for the deaf.

In all these cases the electrical stimulation is used to overcome bad, missing or lost body functions and therefore this technique is called functional electrical stimulation (FES). When FES is restricted to the stimulation of the neuromuscular system, it is often referred to as functional neuromuscular stimulation (FNS).

The right recruitment order of the nerve fibers is very important, e.g., to allow paralyzed patients to move their limbs in a balanced way. On the other hand, deaf people can perceive auditory sensations by electrical stimulation if the auditory nerve is still working, however, speech understanding demands a high coding technology in order to produce the array of nerve signals needed to transmit the information to the brain. Functional electrical stimulation is a young scientific discipline and a high expenditure of research about electrode geometry and signal coding will be necessary for optimally helping the handicapped.

In this book we will be concerned mainly with the basic knowledge for electrical nerve stimulation, used to explain and optimize the firing sequences of stimulated nerves. These depend on the shape and strength of the electrode currents, as well as, on electrode- and body geometry.

But before going into details we will start with historical remarks about the milestones in electrical stimulation and neurophysiology. Detailed technical information on electrodes, tissue damage and physiological considerations are beyond the scope of this book but an introduction to this topic is given in this Chapter and Chapter 2.

Electrical properties of animals were known in ancient Egypt and we have a record dated from about 2600 B.C. of the electric catfish *malapterurus* (FRITSCH, 1887). The medical use of electrical currents has been known for about 2000 years (McNEAL, 1977). In the year 46 one of the first Roman physicians, SCRIBONIUS LARGUS, recommended in his book *Compositiones Medicae* treatments against headache and gout with electric shocks from the torpedo fish. Of course the torpedo fish is not a convenient therapeutic device, and it was not until the middle of the 18th century, when electric shocks were produced with the help of electrostatic machines (Fig. 1.1) and Leyden jars in order to treat paralyzed patients.

In 1791 LUIGI GALVANI demonstrated the electrical stimulation of nerves and muscles by using a bimetallic rod (Fig. 1.2). He assumed that muscle contraction was caused by the discharge of 'animal electricity', but in 1793 VOLTA realized that the source of electricity was the bimetallic rod rather than the animal. An understanding of what is really going on when a nerve is electrically excited is not possible without knowledge of the behavior of the membrane which surrounds the nerve fiber. Therefore many contradictory theories have evolved over the centuries until the appearance of the famous work of HODGKIN and HUXLEY. In 1952 they discovered the process of the excitability of nerve fibers, through current-voltage examinations on giant squid axons with the help of voltage clamp experiments. The membrane currents were measured in controlled voltage steps. [For further reference see Chapter 4.] In 1963 HODGKIN and HUXLEY received, together with ECCLES, the Nobel prize for their pioneering work in neurophysiology.

After the invention of MUSCHENBROECK's 'Leyden jar' in 1745, electrotherapy found a broad field of applications. Several investigators looked for relations between signal strength, and stimulus duration τ, which fit with experimental data. In 1892 the physicist HOORWEG published a formula for the voltage V which is necessary to produce an excitation.

$$V = a \cdot R + \frac{b}{C}$$

V is the voltage to which the capacitor (the Leyden jar) is to be

Fig. 1.1 Electricity was generated in an electrostatic machine (right) and stored in a Leyden jar (arrow) in order to shock patients suffering from paralysis or convulsive fits.

(After G. Adams, 1799)

loaded, R is the resistance of the discharge unit, C is the capacity, and a and b are coefficients determined by the specimen. The following threshold relations, where the charge Q, voltage V, or current I were expressed as functions of duration τ, were found by other early investigators

$Q = a + b\tau$	WEISS (1901)
$V = \frac{a}{1-e^{-\frac{\tau}{b}}}$	LAPICQUE (1907)
$I = \frac{a}{\sqrt{\tau}}$	NERNST (1908)

All these formulas show that the stimulus signal strength de-

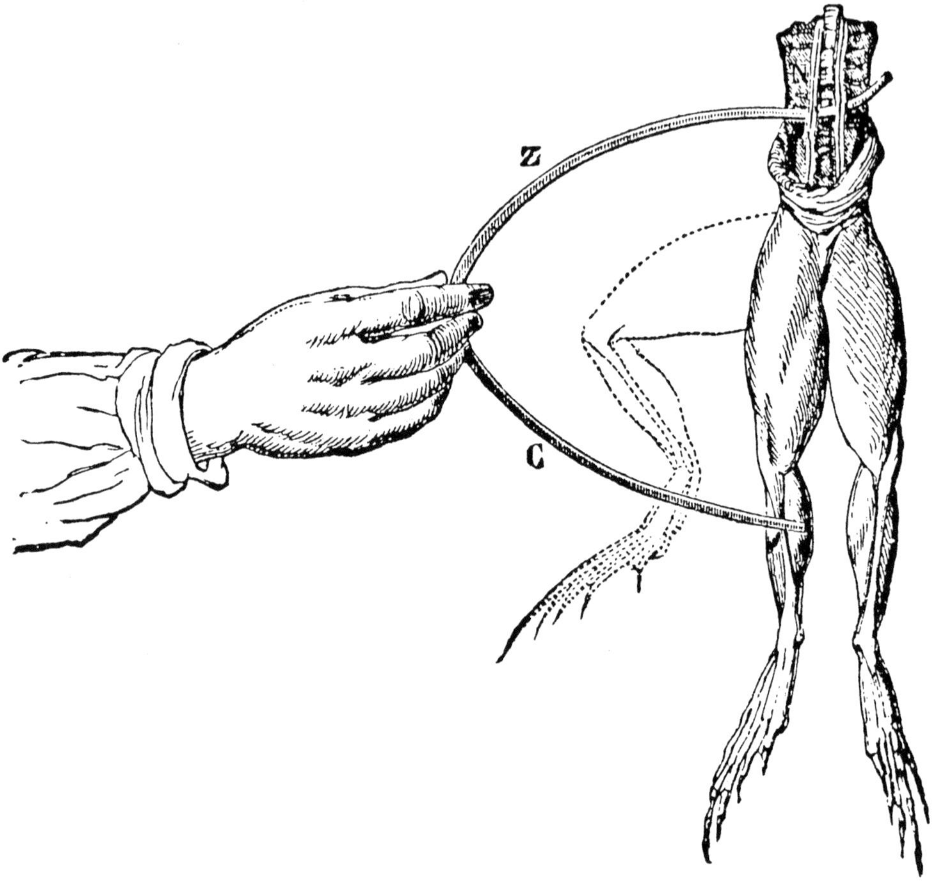

Fig. 1.2 In 1791 and 1793 Galvani and Volta have shown that muscle contraction is possible when a bi-metallic rod touches the nerve or the muscle of a frog's leg.

This experiment demonstrates clearly that functional electrostimulation can overcome interruptions of the neural path from the brain to the target neurons und thus, patients can move their paralized limbs when stimulated electrically.

(From Beard and Rockwell, 1878)

creases with increasing pulse time. The formulation of HOORWEG, WEISS and LAPICQUE demonstrates that even for an infinite application, the signal amplitude does not lie below a certain value - which is in accordance with 'modern' strength-duration curves.

Many electrical devices were in use in the 19th century for different treatments. They were reported to cure rheumatism, neuralgia, fractures, bruises, cuts, insomnia and even cold feet. In 1825 SARLANDIERE treated his patients with electric acupuncture against chronic pain and he claimed that electrical stimulation 'confused' the perception of pain signals. Three years later KRIMER made his first attempts at cardiac resuscitation by electrical stimulation with needles placed directly in the myocardium. His attempts failed but 100 years later the same technique was used successfully by HYMAN.

The modern era of functional electrical stimulation began with the invention of the artificial cardiac pacemaker. In 1952 ZOLL could maintain the heartbeat of a patient who suffered from the Adam - Stokes syndrom (heartbeat had stopped) for twenty minutes, by placing the electrodes of a pacemaker on the chest of the subject. The first successful long-term application of the pacemaker, for 96 days, was achieved by FURMAN and SCHWEDEL (1958), and the first completely implantable pacemaker was installed a few months later by SENNING (October 1958). He used two nickel-cadmium cells as power supply which were charged inductively through the skin. At the same time GLENN *et al.* developed a radio frequency-coupled power supply. In 1960 CHARDACK *et al.* implanted the first pacemaker which required no charging from the outside.

The history of auditory prostheses also goes back to VOLTA. In 1790 he connected his own ears to his newly discovered voltaic cells and he described auditory sensations. In 1957 DJOURNO and EYRIES reported the electrical stimulation of the auditory nerve of a deaf patient. This led to the development of implantable devices by three different groups of investigators in California (SIMMONS, 1966; MICHELSON, 1971; HOUSE 1976).

Scientific application of functional electrical stimulation demands some basic knowledge about the mechanism of natural excitation. Today we know that the nerve system uses action potentials (AP) which propagate along nerve fibers (axons) in order to transmit information. In 1905 HERMANN conceived the idea that impulse propagation is caused by the current flow from an active region which stimulates the resting region ahead. The ALL–OR–NONE CONCEPT was claimed: the propagating AP has a specific strength; if it is to small it cannot propagate over long distances. This rule was proved accurate for muscles by LUCAS (1909), and for nerves by ADRIAN (1914).

KOHLRAUSCH (1876), ARRHENIUS (1887), NERNST (1888, 1889) and BOLTZMANN were concerned with electrolytes and electrolytic conduction, which were soon widely used in the investigations of biological systems. Of great interest is the early hypothesis of BERNSTEIN (1868, 1902, 1912). He hypothesized that the interior of a living cell is an electrolyte where ions move freely - much more than they do on the outside. Furthermore, he assumed that the surface of the cell is covered with a membrane only slightly permeable to potassium ions, but when the membrane becomes activated the selective permeability for potassium is lost and the membrane becomes permeable to all ions.

At that time, the physical structure of a living cell was wholly unknown. The hypothesis of an enveloping membrane was proved by measurements of FRICKE (1923) who found the capacity of red blood cell membranes being 0.81 $\mu F/cm^2$. [Capacities of cell membranes have a value of about 1 $\mu F/cm^2$ with a range from about 0.5 to 2 $\mu F/cm^2$ (COLE, 1968).] Assuming that the membrane is of oily consistence with a dielectric constant of 3, FRICKE obtained a membrane thickness of 3.3 nm, which is a molecular dimension. The conductance of a resting cell membrane was found to be of the order of 1 mS/cm^2 but variation is much higher than in the case of membrane capacity - from 0,01 to 100 mS/cm^2 (COLE, 1968). [More about membranes can be found in Chapter 3.]

Progress in understanding the propagation of nerve signals was possible by experiments on giant squid axons (YOUNG, 1936; COLE & CURTIS, 1939; COLE & HODGKIN, 1939; HODGKIN & HUXLEY, 1939), because the extremely large axon diameter, up to 1 mm, allowed the insertion of microelectrodes available at that time. The breakthrough in understanding the propagation process (without having a clear picture of membrane kinetics) came with the description of membrane behavior by a system of four differential equations which are known as the HODGKIN-HUXLEY equations (HODGKIN & HUXLEY, 1952).

The properties of a biological membrane depend essentially on the types and the number of ionic channels. We find, of course, qualitative differences in the membranes of living cells. The myelinated axons of vertebrates have regions with high activities in the NODES OF RANVIER, whereas very little current runs across the membrane in the internodal area where the membrane is cov-

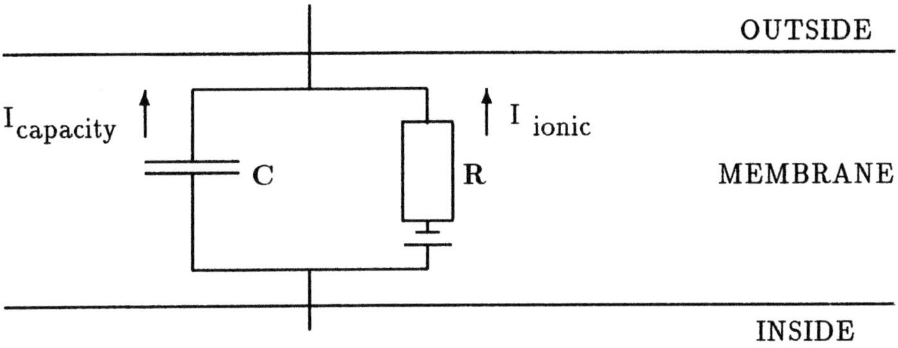

Fig. 1.3 Electric circuit for a patch of membrane. Different ionic concentrations on both sides of the membrane cause a voltage even in the resting state, which is represented as a voltage source in the circuit. Membrane resistance is highly nonlinear because of the activity of the ionic channels embedded in the membrane. For details see Chapter 3.

ered with multiple sheets of the insulating myeline. Technical progress allowed the measurement of myelinated frog axons which have diameters of some μm only. FRANKENHAEUSER & HUXLEY (1964) published their equations which seemed more appropriate for mammalian and human applications.

The HODGKIN-HUXLEY equations and their modified forms as well as the FRANKENHAEUSER-HUXLEY equations, describe quantitatively the voltage current relations of a piece of membrane in the time dimension. The behavior of such a patch of membrane can be simulated by an electric circuit consisting of a voltage source, capacity and nonlinear resistance (Fig. 1.3). If we assume an inside potenial V_i and an external potential V_e we obtain the voltage $V = V_i - V_e$ across the membrane. The current I_m through the membrane is

$$I_m = I_{capacity} + I_{ionic} = C \cdot \frac{dV}{dt} + \frac{V}{R} \qquad (1.1)$$

where C is the capacity of the membrane and R (which is a function of V and time t) is the nonlinear membrane resistance. This is one differential equation. The other three differential equations

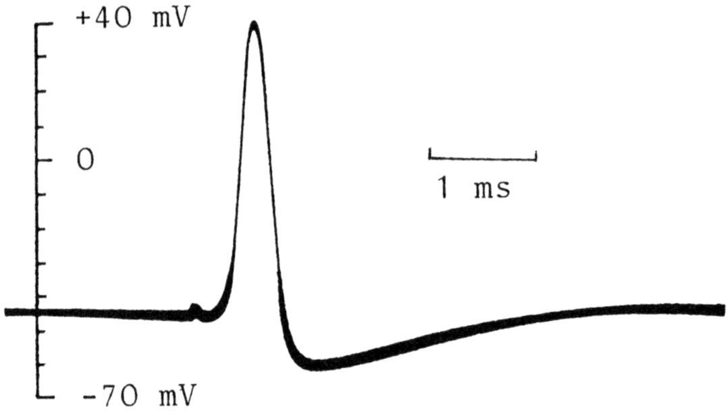

Fig. 1.4 *An action potential from a squid giant axon. The vertical scale indicates membrane voltage in mV. In the resting state the inside potential is negative respective to the outside. The action potential is preceded by a small stimulus artifact. During activity the voltage across the membrane changes polarity and the inside becomes positive. After this depolarization the inside potential becomes more negative than at the begining (hyperpolarization) before it finally returns again to the resting value.*

(After Hodgkin and Huxley, 1945)

formulated by HODGKIN and HUXLEY are used to describe the gating process of the membrane channels in a statistical manner. The whole propagating process of nerve signals is only possible because the conductance of the membrane to ionic currents depends essentially on voltage.

We assume now that the membrane is in the resting state: there is no capacitive current, and the ionic currents are small enough to be neglected. When the system is sufficiently disturbed, e.g., by applying an additional electrode current pulse, an action potential will be generated and the whole process can be described with the HODGKIN-HUXLEY equations. By increasing the voltage, sodium channels open up and the excitation starts with an influx of sodium ions. This happens because there are high outside and low inside concentrations of sodium ions. During excitation the membrane voltage (inside against outside) climbs from about –50 to –70 mV (**polarization**) up to about +40 mV (**depolarization**). Throughout this change in voltage, the potassium channels open and allow the highly concentrated inside potassium ions to

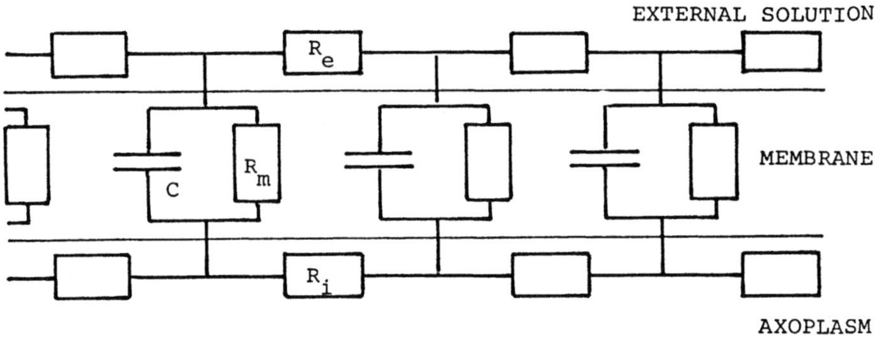

Fig. 1.5 Classical electrical network for the membrane of an axon. Each element of the network consists of a circuit as shown in Fig. 1.3 but the voltage source is neglected here. The circuits are connected via the inneraxonal resistances R_i and the extracellular resistances R_e. If no artificial electrical field is applied the calculations can be simplified by setting the extracellular potential zero because of the large external conducting medium.

flow outside, thereby reducing the membrane voltage. This excitation process is typical for nerve and muscle fibers in all animals (Fig. 1.4).

Because the HODGKIN-HUXLEY equations consist of complicated nonlinearities which are necessary to fit the experimental results, it was a very time consuming task for both investigators to compute the action potentials with the calculators of the early fifties. Nevertheless, the computations were in good agreement with the measurements. [More details are presented in Chapter 4.]

When an axon is sending a signal, the excited region has a length of only a few *cm*, and this excited region propagates along the fiber. We can simulate such a propagating action potential with a ladder network (Fig. 1.5) consisting of circuit elements, of the type shown in Fig. 1.3, where each circuit describes the reaction of a small piece of the axon. [See also Chapter 6.] Such networks have been used to describe the cables for electric telegraphy (KELVIN, 1855) but they were also applied since the turn of the century in electrophysiology (HOORWEG, 1898; HERMANN, 1905; TAYLOR, 1963; COLE, 1968).

For simplicity, the early investigators assumed that the extracellular potential was not influenced by nerve activities and they set $V_e = 0$; membrane voltage V then becomes equal to the inner potential V_i. The network of Fig. 1.5 consists now of the constant elements C (membrane capacity), R_i (inner resistance) and R_e (extracellular resistance) but the nonlinear membrane resistance R_m, membrane current I_m and voltage become functions of t and x, where x is fiber's length coordinate. With these assumptions one can derive a partial differential equation (e.g. COLE, 1968, p. 67-68) for the membrane currents from the network of Fig. 1.5

$$\frac{1}{R_i + R_e} \cdot \frac{\partial^2 V}{\partial x^2} = C \cdot \frac{\partial V}{\partial t} + \frac{V}{R_m} \tag{1.2}$$

This is a diffusion equation which is difficult to evaluate because of the nonlinear membrane resistance R_m. Only for subthreshold reactions can we simplify the calculation by using constant R_m and obtain the 'cable equation' which was first examined by LORD KELVIN (1855) in order to deal with the attenuation and distortion in the first Atlantic cable. [For details see COLE, 1968; SCOTT, 1977; JACK *et al.*, 1983.]

The models of HODGKIN-HUXLEY and FRANKENHAEUSER-HUXLEY use several nonlinear terms to fit the experimental data. Thus, analytical treatment is very difficult and several mathematical investigators tried to simplify the models; but as new variables without physical interpretions were introduced, such models are commonly not of great practical use. More relevant results will be obtained by computer simulations of physical models. The first numerical evaluation of the network shown in Fig. 1.5 (Eqn. 1.2) was done in 1966 by COOLEY & DODGE.

The network of Fig. 1.5 can also be used to simulate electrical nerve stimulation when we do not set $V_e = 0$ but calculate the extracellular potential from electrode currents. [See Chapter 7 and Chapter 8.] As we will see in the later chapters of this book, important results about the influence of geometric parameters can be found by simulation of this network, but it took until 1976, when first computer results were produced in this way by MCNEAL. Ten years later the present author made an analysis of this network and found that the second derivative of the extracellular potential is of essential importance for the extracellular electrical excitation (RATTAY 1986b, 1987b, 1989). [See also Chapter 7.] This is in contrast to older assumptions that the current densities produced by the electrodes in the vincinity of nerve fibers are a measure for stimulation.

Electrical stimulation of nerve and muscle fibers

In the following chapters we will be concerned with the excitability of fibers in detail, but we give here a short summary of the excitation principle.

As a consequence of different ionic concentrations at the inside and the outside of a nerve or muscle fiber, the interior of the cell is normally maintained at a potential of about 50 – 70 mV negative to the exterior. When an action potential is produced the voltage is changed to positive values before it falls back again to the resting state (Fig. 1.4). Such an action potential propagates because it disturbs the resting region ahead by itself and channel activities are evoked there, too. A similar effect can be reached artificially by making the inside potential more positive with the help of an inserted microelectrode as described in Chapter 4. However, placing an electrode inside the cell is usually not practical for clinical applications; therefore, nerve and muscle fibers are usually stimulated by changing the membrane voltage via the extracellular potential.

The choice of electrodes

In order to stimulate nerve fibers artificially an electric field must be created. For this purpose, neural prostheses use either surface electrodes or implanted electrodes*. Both types of electrodes have their applications (VOSSIUS 1983, 1986).

Surface electrodes need no surgery, but the large distances to the stimulated areas and the insulation of the skin and fat areas demand high stimulation strengths with low fiber selectivity. To overcome the high resistance of dry skin, they may consist of a pad soaked with sodium chloride, a carbon donated rubber or silicon pad, or a special metal buffered with an electrode jelly. Because the skin is irritated by the chemical substances which are used for good conductance, surface electrodes are removed when not in use. When the electrodes are used again, inexact positioning will

* An alternative technique is the noninvasive stimulation with coils, which produce extracellular potential distributions by time varying magnetic fields. See Chapter 11.

cause problems because other groups of fibers are stimulated or a variation of the current strength would be necessary as a consequence of changed distances. Special garments, where electrodes are sewn on in well-defined positions put the equipment quickly into operation.

Surface electrodes might be preferred in functional neuromuscular stimulation under the following circumstances:

- if one starts with stimulation rather early after an injury, the extent of paralysis might change especially after the first phase, thus requiring relocation of the electrodes more often;
- during the training period of the muscles, the muscle-volume might increase drastically. In general large areas may be covered with surface electrodes which facilitate exercising;
- in special cases of spasticity in order to built up the antagonistic muscles;
- in order to bring about simple movements which require only few electrodes with easy positioning.

Implanted electrodes demand high-quality materials because the surface should not be changed electrolytically. If the electrodes are used for neuromuscular stimulation they consist mostly of very thin coiled stainless steel wires which move with the muscles and other tissues without breakage due to mechanical stress. In order to hold their position, the electrodes are often fixed in a cuff which surrounds the nerve or they are sewn on at the epineurum. The great advantage of implanted electrodes is that they can be positioned very close to the stimulated fiber. This allows much lower stimulation signals and better fiber selectivity than that possible with surface electrodes.

In the case of neuromuscular stimulation, implanted electrodes are preferred under the following circumstances:

- when deep muscles difficult to excite by surface electrodes, have to be stimulated;
- for complex movements which require a large number of electrodes in a limited space. As an example, more than six surface electrodes cannot be fixed, e.g., on the forearm in order to stimulate complex hand operations;
- to avoid pain sensations, a disadvantage of surface electrodes, because implanted electrodes do not stimulate the skin's pain-receptors.

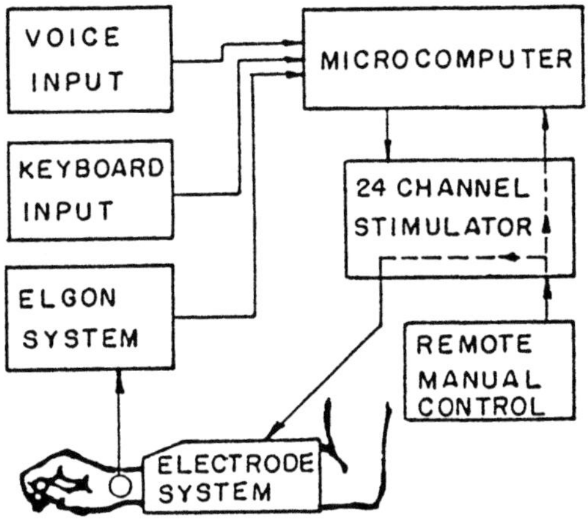

Fig. 1.6 *System components for the neuroelectrical stimulation of nine individual muscles of the forearm and hand by surface electrodes. A microcomputer receives the movement commands by voice or keyboard inputs. Additionally, an electrogoniometer can measure flexion-extension and other feedback data in order to supply the microcomputer with the necessary information for the coding strategy. A square wave double pulse with variable pulse width, frequency and amplitude is thus generated in a 24 channel stimulator. Conductive rubber surface electrodes are positioned on the forearm and on the hand. The electrodes are placed over the respective muscle in such a way that the response is maximal and isolated (no excitation of other muscles). Three types of gripping were programmed: grasp, pinch grip and key grip.*

(From Nathan, 1986)

A typical application where surface electrodes as well as implanted electrodes can be used is to restore the handfunction. The fingers have to be opened and closed to perform different types of grips according to the task to be executed. With implanted electrodes the movements of the fingers can be stimulated more precisely, but for practical use the difference is not conspicuous.

NATHAN (1990) describes a multichannel system, combined with surface electrodes, which allows a variety of functional arm movements to be restored to C4 spinal lesion quadriplegic subjects (totally paralyzed in both arms) including eating, drinking, hair- and tooth-brushing, application of makeup and even writing.

The principal technical problem is targeting the muscles. Up to 12 individual muscles are used for finger, thumb and wrist joint movements and elbow joint articulations in the Beer Sheva system. A special technique has been developed in order to find the motor points, which are the positions on the skin where two bipolar surface electrodes produce maximum strength of contraction in the target muscle with minimum side effects to other muscles.

Fig. 1.6 shows schematically that such a stimulation system can be used to restore handfunction with the help of surface electrodes, but the whole system is typical also for implanted electrodes and can be generally applied in neuromuscular stimulation. [For more details see Chapter 10; NATHAN, 1986]

Electrode processes and tissue damage during stimulation - monophasic and biphasic signals

If implanted electrodes are in use, attention should be paid to the right electrode material as well as to the strength and the form of the stimulus signal in order to prevent damage to tissue and electrode. Electrical nerve stimulation is a consequence of electron migration in the electrode and of ion migration in the tissue medium. At the electrode-tissue interface of implanted electrodes some processes must occur to support the conversion of the charge carrier (LOEB *et al.*, 1982). They can change the composition of the chemical species in the immediate vicinity of the stimulating electrodes. As an example, Fig. 1.7 illustrates the processes involved with a platin electrode when a strong biphasic stimulus signal is applied.

For medical applications these processes should be reversible, i.e., if new chemical species are involved they should remain bound to the electrode surface. Otherwise some of the products will escape from the electrode. Irreversible charge injection can lead to cumulative chemical changes which include water electrolysis to H_2 and O_2, saline oxidation, metal dissolution and oxidation of organic molecules. Changes in pH can interrupt local blood flow or cause irreversible changes in tissue proteins. Some of the metal ions, e.g., cobalt, silver, lead and mercury are toxic to neural tissue.

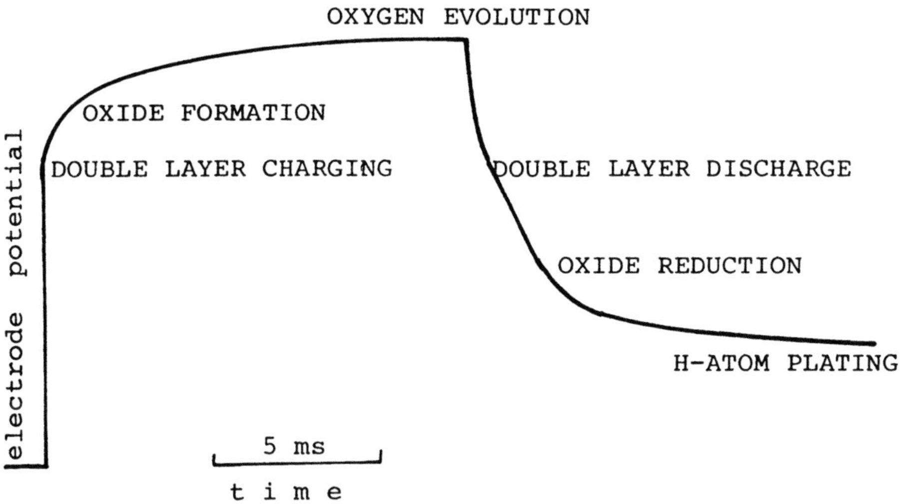

Fig. 1.7 *The main processes and the potential transient produced by a single biphasic pulse to a platin electrode in a bath solution when stimulated with an anodic first pulse. (500 μC per cm² of electrode surface; 32 mA/cm²).*

A layer of oriented water molecules at the electrode surface acts as a dielectric and the whole system works like a condenser with a capacity of 10-20 μF/cm². Useful charge injection is limited to a few μC/cm² before faradaic reactions begin. These involve electron transfer across the interface resulting in the formation of new chemical species. This picture shows Pt-reaction occuring when strong stimuli are applied. The curve is divided into several regions according to the main processes.

The anodic charge injection starts with the capacitive charging of the double layer (condenser). After the injection of about 5 μC/cm² oxide formation becomes dominant. The next phase is characterized by oxygen evolution, followed by a small region of double layer charging caused by the reversal of the current. Afterwards Pt oxide is reduced in a large region and finally H-atom plating occurs.

If we start at first with a cathodic pulse the main processes would be: double layer charging - H-atom plating - H_2 evolution - reoxidation of H-atoms - double layer charging - oxide formation.

(After Brummer et al., 1983)

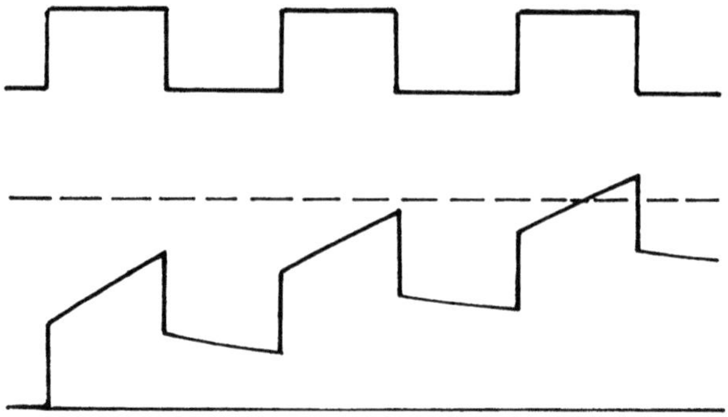

Fig. 1.8 *Stimulation with a train of monophasic current pulses (upper trace) can be dangerous because of the accumulating effects of charge densities and electrode potential (lower trace) even if the application of one single pulse would not be critical. The critical charge is assumed to be at the dashed line.*

But even without applying an electric field metal ions may migrate away from the metal surface into the surrounding solution. This causes a potential difference which in turn provides a force of attraction for ions in the solution and results in the corrision of the electrodes.

Nevertheless, nerve and muscle fibers can be stimulated safely for prolonged periods of time if care is taken when choosing the electrode material and the stimulus parameters. In order to avoid irreversible processes, the charge density at the surface of the electrode and, respectively, the electrode potential has to be kept within certain limits.

For reasons of safety, biphasic stimulus signals are preferred for most applications since charge accumulation can thus be avoided. Fig. 1.8 shows schematically the effect of the accumulation of electrode potential when stimulation occurs with monophasic current pulses. The current always has the same polarity. Accumulation of electrode potential and of charge density by series of monophasic current pulses can lead into a critical region.

High charge densities can be avoided by using trains of biphasic impulses (Fig. 1.9). The function of the primary pulse is to produce an action potential and the second pulse is used to reverse the electrochemical process. The second pulse has a hyperpolar-

Fig. 1.9 Trains of biphasic current signals are preferred in stimulations because they allow the application of balanced charge densities (compare also Fig. 9.6). (A) periodic biphasic impulses. (B) biphasic impulses with a gap between the pulses need smaller threshold amplitudes than the signals showed in (A) especially when short pulses (shorter than 200μs) are applied. (C) A signal with an exponential decay in the second pulse is often used because it is technically easy to realize. See Fig. 1.10.

izing effect and reduces the stimulating work of the first pulse especially when it is applied shortly after the onset of the first pulse. A delay between these pulses is often used because it will reduce this effect. [See also chapter 9.] Fig. 1.9 (C) and Fig. 1.10 show a signal form with exponential decay which is often in use in order to avoid charge accumulation.

Future developments

One of the main problems in functional electrical stimulation is to stimulate the nerve fibers individually according to a special code. In Chapter 10 we will discuss different techniques which are being developed in order to imitate the natural recruitment

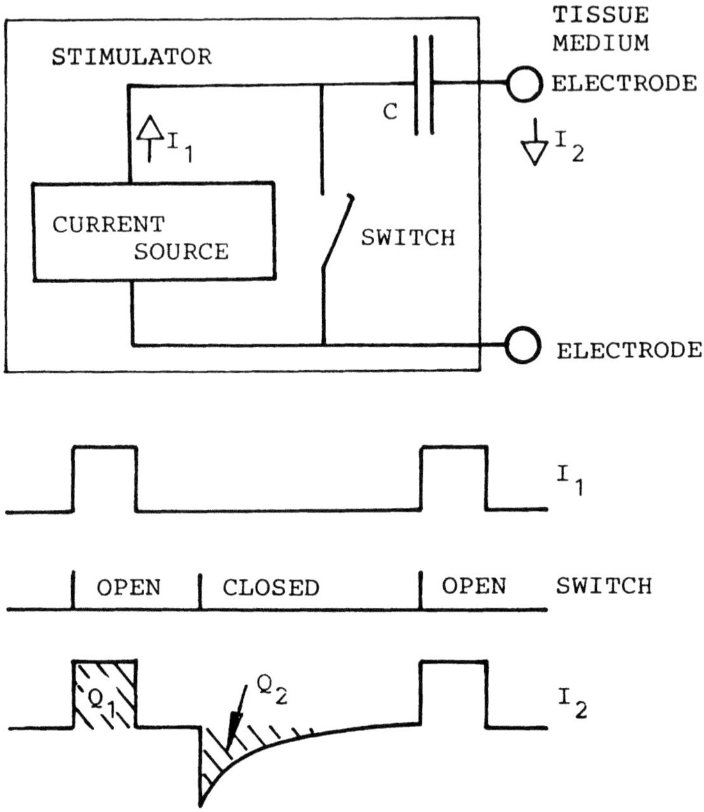

Fig. 1.10 *A circuit with a condenser and a switch can be used to produce charge balanced stimulus signals. The charge injected during the primary pulse is stored in the capacitor C while the electronic switch is open. Closing the switch allows to discharge C. As $Q_1 = Q_2$ there is no charge accumulation at the electrodes.* *(After Mortimer, 1981)*

order of motor neurons. Another task is to supply the sensory nerves (of handicapped people) by electrical stimulation with the right information. In Chapter 12 the complexity of the array of spikes which transforms the information in the acoustic nerve will be discussed. Today we do not even know the neural pattern which codes simultaneously the loudness and the frequencies of the acoustic signals; neither we are able to produce the right sequential order individually in the different fibers of the nerve. The speech understanding of deaf patients which are using cochlear prostheses is not very good today, but it is surprisingly good if we take into account our restricted knowledge of the neural coding and the limited possibilities to stimulate the 30,000 fibers of the acoustic nerve individually. We find an even more problematic situation when stimulating the blind electrically.

Several groups have investigated the stimulation with electrodes placed on the surface of the cortex to develop a prothesis for the blind that might be of value in reading and mobility. When the visual cortex is stimulated electrically, human subjects see circumscribed and often punctate sensations of light in different colours (BRINDLEY & LEWIN, 1968; DOBELLE & MLADEJOVSKY, 1974; DOBELLE *et al.*, 1976).

The stimulation becomes more sensitive when intracortical microelectrodes are used. First results are obtained on primates (BARTLETT & DOTY, 1980) and on human subjects (BAK *et al.*, 1990).

The neural code generated by optical signals is more complicated then that caused by accoustic stimulation. Although important findings on the brain mechanisms of vision give more insight (e.g., HUBEL, 1982; HUBEL & WIESEL, 1979), we are still far away from knowing how to stimulate a special area in the visual cortex optimally by an array of electrodes. Thus, great effort over years will be necessary to produce a device for the blind that can generate sensations with a quality comparable to that of the cochlear implants available today.

A first step of individual stimulation of nerve fibers was done with experiments on rats and monkeys. Interfaces between separated fibers and electrodes show good tolerance and functionality. Fig. 1.11 shows the principle of an implant designed by ETTINGER *et al.* in 1987 for the observation of single axon activities. The device can also be used for stimulation. Apparently, the slots have widths of 80 μm to 250 μm, but the implant must be refined if more fibers are to be coupled individually.

Another interesting project is the two-dimensional microelectrode array designed by KOVACS *et al.* in 1987. The implant consists of a plate with an array of laser drilled holes with diameters down to 8 μm. The array should be interposed between the

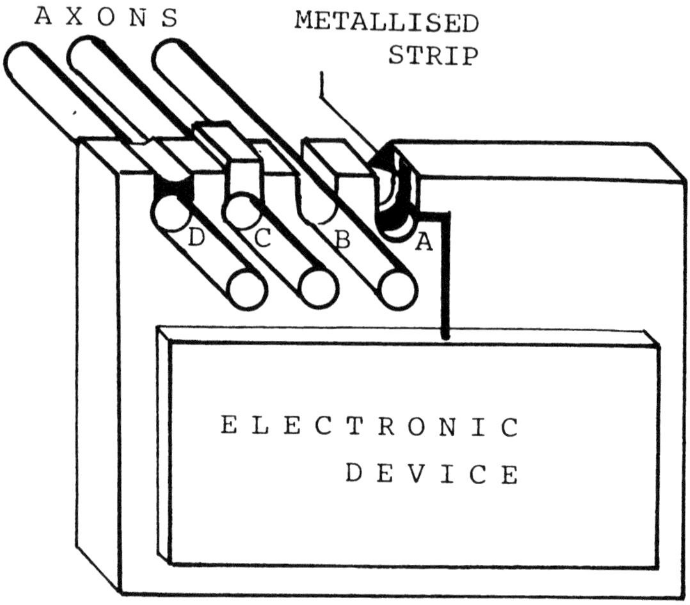

Fig. 1.11 *Microelectronic implant for bidirectional neural interface (schematic). Every slot is used for direct coupling several fibers or only a single fiber with one electrode. The metallic strip (A) in the slot is connected with the microelectronic device and can be used either to observe the occurrence of spikes in the inserted fibers or to stimulate the fibers artificially. After laying in the fiber (B) the contact with the electrode is made by using a plug (C) or adhesives (D). The length of the slots is some millimeters in order to allow close contact of the electrode to a node of the myelinated fiber.*

(After Ettinger et al., 1987)

ends of severed peripheral nerves. The holes allow the axon to regrow through the chip. The growing of the nerve can be influenced by the extracellular components of the environment and the regenerating proximal stump can improve axonal organization (FREED *et al.*, 1985). Although for clinical use, investigations are needed on the nerve growth through the chip, KOVACS *et al.* 1987 demonstrated in 1987 through several examples, that monkey axons grow through the holes of such an implant. Further research has to be done in order to drill the fine holes and to fabricate chips and electrodes with the necessary refinement.

When these developments are successful, a well operating neural interface for individual fiber stimulation and observation will show many new applications in functional electrical stimulation in the next decades. However, using, the state of the art, we have to be concerned mainly with the stimulation of fibers via electric fields produced artificially by electrodes implanted close to the nerve or by surface electrodes. The purpose of this book is to serve as an introduction to the basic knowledge of the subject matter.

2. FUNCTIONAL DESIGN OF THE NERVOUS SYSTEM

The elements of the neuron

The nervous sytem is concerned with the rapid transfer of information through the body in the form of electrical signals. The cells of the nervous system are called **neurons**. In some primitive invertebrates we find a simple network of undifferentiated nerve cells. In contrast, the human neurons show great variations in their shapes in relation to their specialized functions. The bulk of nerve cell bodies are collected in the central nervous system (CNS). The peripheral nerve system consists of the **sensory neurons** (afferent neurons, i.e., those which convey information to the central nervous system) and **motor neurons** or motoneurons (efferent neurons, which convey information from the CNS to the body).

Each neuron has a cell body (**soma**) and one or more processes or extensions. An *unipolar neuron* has only one process, a *bipolar* has two and a *multipolar neuron* has three or more processes. One of the processes is usually much longer than the rest. This is the nerve fiber or **axon** while each of the shorter processes is called **dentrite**. The axon is used to carry the outgoing information into areas, which are in man, up to one meter away from the cell body. At the end region the axon normally branches into several fibers. A widening in the terminal region of the axon is called a **synapse**. At synapses the neuron is in close contact with other cells, and the neural information crosses via the small synaptic cleft into other neurons, muscle fibers or glands.

The neurons of different parts of the nervous system show a wide range of shapes and sizes (Fig. 2.1). For example, the human cerebellum has more than 10^{10} nerve cells which belong to five main types. Besides the Purkinje cells [Fig. 2.1 (A)], whose axons constitute the only output from the cerebellum, we also find basket cells, Golgi cells, granule cells and stellate cells. As illustrated by Fig. 2.1, these different types of neurons have been known for about 100 years.

Fig. 2.2 shows the diagram of a typical neuron: The dentrites, as well as, the cell body are covered with up to 200,000

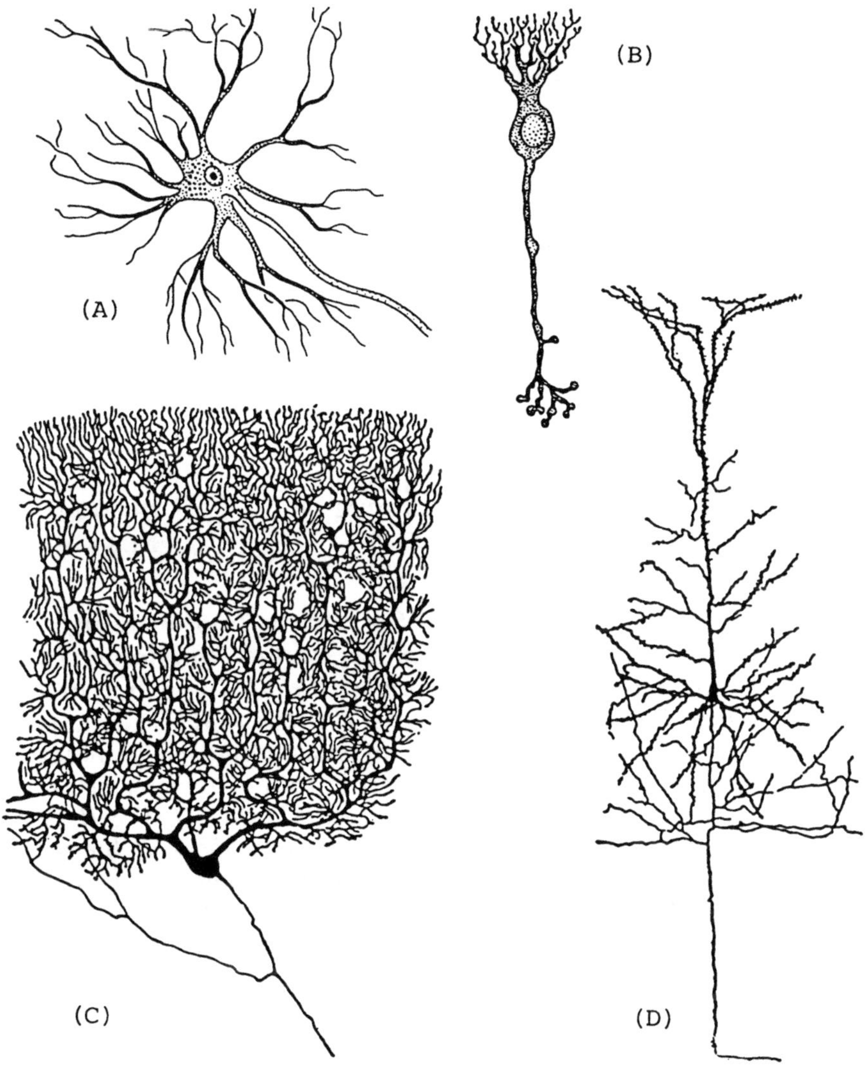

Fig. 2.1 Some typical shapes of neurons. *(A) Motoneuron dissected from a mammalian spinal cord. (B) Bipolar cell from the retina of a dog. (C) Purkinje cell from the human cerebellum. (D) Pyramidal cell from the cerebral cortex of a rabbit. In the cases (A), (C) and (D) the relatively long axons are not shown completely but they are cut at the right lower edges of the pictures. Thus only the branching of the axon at the terminal region with the synaptic knobs can only be seen in (B).*

(After drawings of Deiters, 1869 and Ramón y Cajal, 1909.)

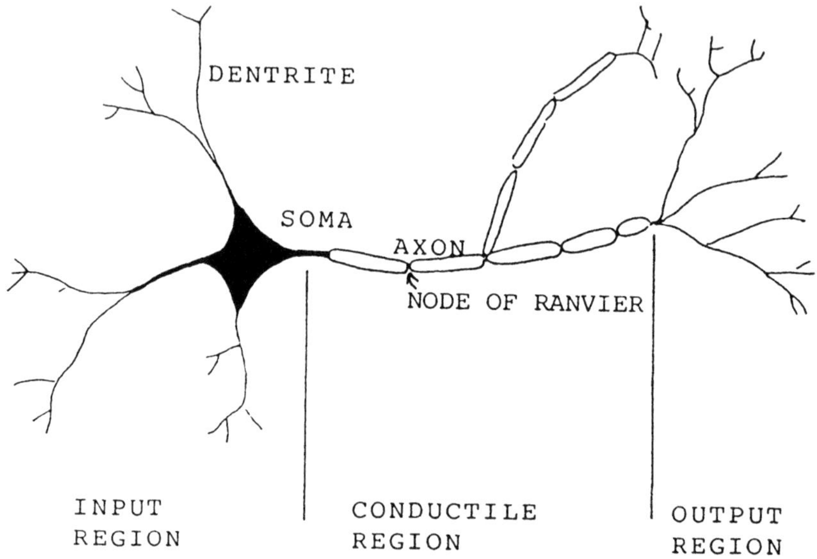

Fig. 2.2 Functional scheme of a neuron (motoneuron). Excitation of the neuron is a consequence of the stimulation by activated synapses in the input region at the dentritic tree and the soma. When the excitation exceeds the threshold value, a single nerve impulse or even a train of such impulses is generated which propagate along the axon into the output regions. Here the synapses help to transmit the information into other cells. The task of the motoneuron is to stimulate muscle fibers at the neuromuscular junction and this is done by special synapses in the terminal region of the axons. Most neurons have much longer axons as shown here. Beside the main axon there can exist a 'collateral branch'.

synapses, bringing the information of other neurons into the nerve cell. Some of the synapses are exhibitory, which means that they promote excitation of the cell when they are activated by their neuron; others are inhibitory. Sensory neurons will be excited by stimulating the receptor regions at the nerve terminals, e.g., by pressure, heat or chemicals. Neural activities at the input region causes fluctuations of the soma inside potential. Such disturbances propagate into the axon, but if they are small, no reaction will result in the terminal region; only excitation above a certain strength (**threshold**) will generate an impulse which propagates along the whole axon, and thus the output region will receive either a complete impulse or nothing. This is known as the **all-or-nothing law.**

In close contact to the neurons are the *glial cells*, which occupy most of the space between the neurons in the central nervous system. Their physiological role is not completely clear today. We know that they do not carry impulses, but their inside potential changes with the activity of the surrounding axons.

The non-myelinated fiber

All nerve fibers consist of a relative long cylinder of *axoplasm* which is surrounded by an electrically excitable membrane. We will see in the following chapters that electrical impulses travel along an axon (which is stimulated at one end) as a consequence of the special properties of the nerve membrane. The conducting velocity depends essentially on fiber diameter and on insulation by sheaths of myelin.

Axons that are not surrounded by several layers of myelin are called non-myelinated or unmyelinated fibers. It was found experimentally, as well as, by theory that the velocity of nerve impulses in non-myelinated axons is proportionate to the square root of the diameter. Since almost all invertebrates are equipped exclusively with non-myelinated fibers, their single fiber diameters have been developed up to one *mm* in cases where rapid conduction is called for. The giant nerve fiber of *squid* becomes very useful for the exploration of the mechanism of propagation of nerve impulses by HODGKIN & HUXLEY (1952) and other investigators, because their large diameters allow the insertion of axial electrodes (Chapter 4).

Also in vertebrates, the small slowly conducting fibers (below 1 μm) are unmyelinated. To the group of unmyelinated axons belongs most fibers of the autonomic system (innervating viscera), as well as, peripheral sensory fibers subserving sensations like pain and temperature where a rapid response is not required.

It is somewhat contradictory, that unmyelinated fibers sometimes are partly covered with the myelin of special cells, which are called *Schwann cells* at peripheral nerves. Some of the fibers are deeply embedded within the Schwann cell, others are almost uncovered, and in general, each Schwann cell supports a small group of axons. But in all non-myelinated nerves, the axon membrane is separated from the Schwann cells by a space of about 10 *nm* (mesaxon) which is connected with the extracellular space of the tissue. Therefore, ionic currents can pass through the membrane at the whole length of the fiber and the partial insulation of the Schwann cells will not essentially influence the propagating process.

The myelinated fiber

In the vertebrate nervous system, the larger fibers (greater 1-2 μm) are myelinated. Myelin is formed by Schwann cells, which wrap themselves tighty around the axon. With each wrap (*lamellae*) the Schwann cell cytoplasm, which is originally enclosed between two layers of cell membrane, is squeezed out and thus finally a spiral of tightly packed membranes (myelin) covers the axon. The number of wrappings varies from about 10 up to 160 (ARBUTHNOTT *et al.*, 1980). In this way the electrical insulation of the axon is considerably higher than in a non-myelinated fiber, because instead of one membrane, there are now up to 321 layers of membrane between the inside of the fiber and the extracellular space. A double layer of Schwann cell membrane was determined by X-ray diffraction to be about 17 *nm* thick. As an example, a nerve fibre with an outside diameter of 10 μm may have a myelin sheath of 1.3 μm consisting of 75 double layers of cell membrane, thus the inner fiber diameter becomes 74 % of the outer one. Generally speaking, the myelin occupies 20-40 % of the overall fiber diameter.

The myelin sheath is interrupted periodically by 1-2.5 μm gaps called NODES OF RANVIER, where the axonal membrane is exposed. The internodal distance (ranging from 0.2 to 2 *mm*) is usually about 100 times the diameter of the outer fiber. As an effect of myelination, the activity of the membrane is restricted to the free regions at the nodes of Ranvier, because the ionic currents cannot pass through the many layers of Schwann cells. The capacitive currents are also very small because the capacity of a serie of N equal condensors is only the N-th part of the capacity of a single condensor.

In contrast to the **quadratic relation between fiber diameter and propagation velocity for unmyelinated axons,** the dependance becomes approximately linear in the case of myelinated fibers [Fig. 6.7; RUSHTON, 1951 (theoretical) and ARBUTHNOTT *et al.*, 1980 (experimental)].

In practice, this means that for warm blooded **myelinated fibers having diameters less than 11 μm, the conduction velocity in meters per second can be assumed to be 4.5 times the diameter in** μm, and for thicker fibers this constant of proportionality becomes about 6.

Branching of dentrites and axons

In the conduction of myelinated peripheral nerve fibers, there is a safety factor of about five (KUFFLER *et al.*, 1984, p. 183). This means that the depolarization produced at a node, due to the excitation of the preceding node, is approximately five times larger than necessary to reach threshold. This safety factor can be considerably reduced in morphological circumstances. In the case when an axon branches, the current supplying the last single node at the branch is divided between two nodes beyond the branch. The safety factor for conduction at one or both branches may be reduced, and this reduction depends on geometry (diameters and numbers of branches).

SCOTT (1977) reviews theoretical work on the passage of action potentials in fiber areas with changing diameters and at branches. It was shown by simulation of squid axons that the abrupt widening of fiber diameter will block a single action potential. For example, at a five-fold increase of fiber diameter, an action potential will pass the discontinuity with considerable delay; however, with a six-fold increase conduction is blocked (KHODOROV *et al.*, 1969). The blocking effect of the widened fiber occurs even for smaller ratios of diameter when a pulse train is applied. This was demonstrated by KHODOROV *et al.* (1971) theoretically and the result is illustrated in Table 2.1.

It was shown that marginal passage of a pulse, caused by a widening, may lead to time delays in the order of 0.5 - 1 *ms* (KHODOROV, 1974). This may be of importance when the dentrite enters the expansion the soma, e.g., in the bipolar cells of the auditory nerve these enlargements are probably an essential factor having influence on the pulse train coding (SCOTT, 1977, p.133; BOGOSLOVSKAYA *et al.* 1973).

Table 2.1

Blocking of a periodic train of impulses caused by abrupt widening of a non-myelinated fiber

abrupt widening ratio	interpulse interval (msec) 2.5	3	3.3	3.5
	Number of blocked pulses			
3:2	none	none	none	none
3:1	2,4,6, ...	3,5,7, ...	5,9, ...	
5:1	— all pulses except the first are blocked —			
6:1	all	all	all	all

(After Khodorov, 1974)

Fig. 2.3 illustrates some effects caused by an abrupt widening in a fiber which is stimulated by a train of injected current pulses. Every line of the picture belongs to a fiber region of 1 mm length. The first line shows the reaction of the fiber at the top, where from the tip of an electrode, current pulses are applied intracellularly. The stimulus signal is strong enough to produce a synchronized membrane voltage which increases about 100 mV above the resting value (top line), but the firing frequency is reduced twice on the way to the bottom of the fiber (bottom line). The first loss of signals is caused by the long refractory period of the fiber. When the second stimulating impulse is applied 4 ms after the first one, the stimulated fiber is not able to react with a second spike (as seen from line two).

Although there was no noise signal involved in the simulation, the reaction of the second segment demonstrates some irregularity in the firing sequence of the fiber: the stimulus pulses with the numbers two, four, six, seven and eight will not propagate along the small part of the fiber, whereas the first, third, and fifth pulse travel with nearly constant velocity, downwards. At the point of expension, a second reduction of firing frequency occurs. Two facts are responsible for this behavior: first it needs more energy to produce an action potential in the widened axon;* and second the fiber is not in the resting state when the second spike arrives

* The excitation of a following segment is caused by the activity of the previous part. As soon as enough energy is deposed in the new

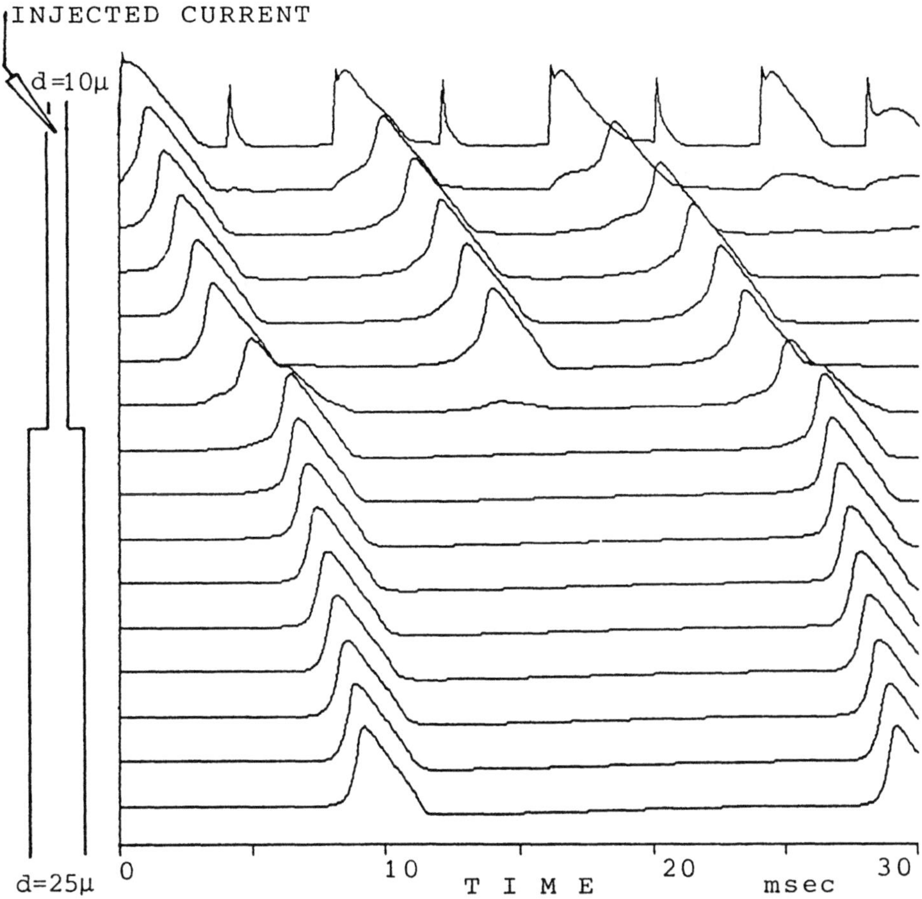

Fig. 2.3 Propagation of excitation along a non-myelinated fiber, which widens abruptly from a 10 μm diameter to 25 μm. On the left side, the form of a fiber is shown which is stimulated by a train of impulses entering the axon intracellularly at the top. In order to compute the reaction, the axon is divided into 1 mm segments. Every line of the right part of the figure shows the excitation of the corresponding segment as a function of time. The first action potential propagates downwards along the fiber. Between the seventh and the eighth segment there is an abrupt increase of the diameter from 10μm to 25 μm which influences the propagation (for further explanation see text).

Calculation was done with the original Hodgkin-Huxley data as given in Chapter 4, but for a changed diameter. Stimulation was with positive (injected) current pulses of 100 μs duration and 4 ms separation.

---

segment the excitation process is started.

approximately 8 ms after the first one at the point of widening.**

On the axonal side, REVENKO *et al.* (1973) investigated the propagation of an action potential from the myelinated region of an axon into the non-myelinated terminal section. In this region the safety factor is low because the exposed fraction of the membrane jumps from a low value in the myelinated region to unity in the uncovered terminal section.*** If channel density is unchanged, a narrowing of fiber diameter, by a factor of about three, is required to ensure conduction; however, even in this case, frequency reduction must be expected in a similar way as shown in Table 2.1.

** By inspection of, e.g., the lowest line of Fig. 2.3 it is seen that the action potential shows after-oscillations with a duration much longer than the action potential itself. At the end of the spike the membrane voltage is lower than in the resting state (hyperpolarization) and therefore it needs more energy to activate the fiber again. The energy brought into the part of widening, by the incoming spike, is just enough to produce excitation in the resting state, but it is not enough to evoke an action potential when the fiber is hyperpolarized.

*** It is perhaps for this reason that the last few internodes before the unmyelinated terminal are shorter than before. (QUICK *et al.*, 1979)

3. THE EXCITABILITY OF CELLS

Bernstein's cell membrane concept

In 1868 BERNSTEIN developed the hypothesis that living cells are composed of an electrolytic interior surrounded by a thin membrane relatively impermeable to ions. Different ionic concentrations on both sides of the membrane cause a difference in electrical potential or 'voltage'. While at rest, the inside of the cell is about $70mV$ more negative than the extracellular fluid. This state is called '**polarized**'. Whenever nerve cells or muscle fibers are active the voltage across the membrane changes from this '**resting voltage**' of $-70mV$ up to $+50mV$ - they become '**depolarized**' for a short time. The voltage comes down again (**repolarization**), commonly with an overshoot to values less than $-70mV$, which is called '**hyperpolarization**'.

Table 3.1 Ionic concentrations in mMol/l

	outside	inside	
Na^+	120	9.2	frog muscle
K^+	2.5	140	
Cl^-	120	(3-4)	
Na^+	460	50	squid axon
K^+	10	400	
Cl^-	540	40-100	
$Isethionate^-$		270	
$Asparate^-$		75	

(After Katz, 1966)

Cell membranes consist of bimolecular lipid layers, where proteins are embedded (BRETSCHER & RAFF, 1975). Each of the phospholipid molecules have two hydrocarbon chains which build up the hydrophobic center of the cell membrane (Fig. 3.1). As a consequence of the molecular structure, the lipid bilayer is mechanically stable, and even when a microelectrode is carefully introduced across the membrane, the gap is filled and ionic separation is not disturbed.

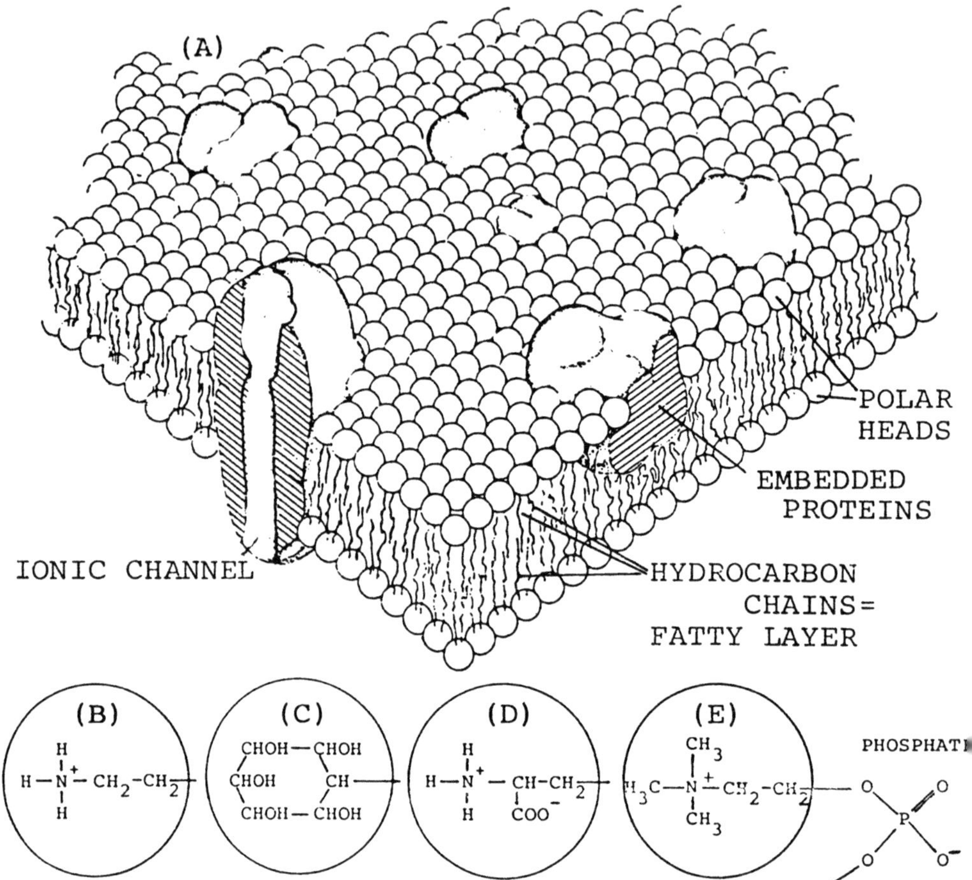

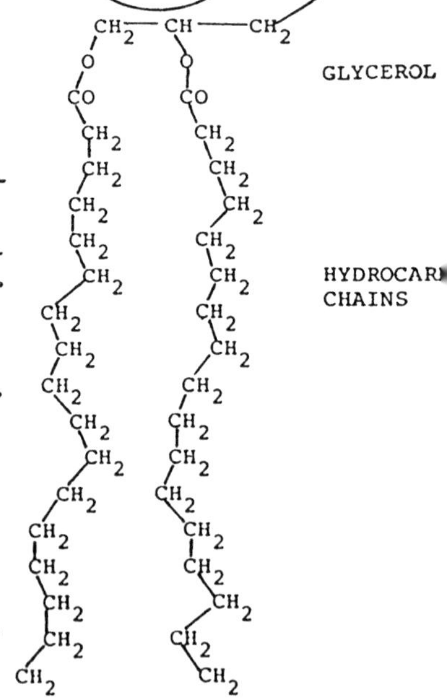

Fig. 3.1 *Model of the cell membrane. (A) Mammalian cell membranes consist of two layers of four different phospholipid molecules: (B) phosphatidylethanolamine, (C) phosphatidylinositol, (D) phosphatidylserine, (E) phosphatidylcholine, which face one another with their fatty (hydrophobic) acid chains containing 18 carbon atoms. Differences between the phospholipids exist only in their polar hydrophilic heads (N^+ in B, D, E and O^- in all cases).*
Small channels intersect the embedded protein molecules and a gating mechanism controls the passage of ions. (After Singer and Nicolson, 1975 and Bretscher, 1985)

BOX 3.1 NERNST- AND GOLDMAN EQUATIONS

When only one type of ions is involved the NERNST EQUATION describes the voltage across the membrane. This results from different ionic concentrations on both sides of the membrane. The GOLDMAN EQUATION can be used for mixed ionic concentrations.

NERNST's basic idea was that the electric work needed to bring n moles of ions from concentration c_1 to concentration c_2 is equal to the osmotic work to compress those ions from volume V_1 to V_2. The ions in an aqueous solution obey the laws of gasdynamics.

In order to compress the volume dV we need the energy $dW = p\ dV$ where p is the pressure. The total work is therefore

$$W_{osmotic} = -\int_{V_1}^{V_2} p\ dV$$

Setting $pV{=}nRT$ with the gas constant $R{=}8.31441J/(mol.K)$ and the absolute temperature T, we get

$$W_{osmotic} = -\int_{V_1}^{V_2} \frac{nRT}{V}\, dV = -nRT \cdot ln\frac{V_2}{V_1}$$

We introduce the concentrations $c_1 = n/V_1$ and $c_2 = n/V_2$. Thus, the osmotic work to bring up n mols from c_1 to c_2 is finally

$$W_{osmotic} = nRT \cdot ln\frac{c_2}{c_1}$$

One mol of ions consists of $6.0225 \cdot 10^{23}$ molecules. The electric work to move the charge Q against the voltage E is

$$W_{electric} = Q \cdot E$$

The Faraday constant $F = 9.64845 \cdot 10^4 C/mol$ gives the charge of one mol of single valenced ions.

With
$$Q = n \cdot z \cdot F,$$
where z is the valence (for Na^+ions, z=1), we get
$$W_{electric} = n.z.F.E$$
Setting $W_{electric} = W_{osmotic}$ results in the NERNST EQUATION for the voltage across the membrane E_m
$$E_m = \frac{R \cdot T}{z \cdot F} ln \frac{c_2}{c_1}$$
Note: At room temperature ($T = 293^o K$) the factor $\frac{RT}{F}$ is about $25mV$.

In many cases, different concentrations of sodium, potassium and cloride ions determine the voltage across the membrane. In 1943 GOLDMAN postulated the 'constant field theory'. The theory assumes:

i) that ions move in the membrane under the influence of electric fields and concentration gradients, just as they do in free solutions;

ii) the ionic concentrations at the edges of the membrane are proportional to those in the aqueous solutions in contact there;

iii) the electrical potential gradient is constant within the membrane.

With these assumptions the GOLDMAN EQUATION allows us to calculate the membrane voltage:
$$E_m = \frac{RT}{F} \cdot ln \frac{P_K[K]_o + P_{Na}[Na]_o + P_{Cl}[Cl]_i}{P_K[K]_i + P_{Na}[Na]_i + P_{Cl}[Cl]_o}$$
where $[K]$ is the potassium concentration and the suffixes i and o stand for inside and outside. P_K, P_{Na} and P_{Cl} are permeabilities measured in $[cm/sec]$. They are defined as $u\beta RT/\alpha F$ where u is the mobility of the ion in the membrane, β is the partition coefficient between the membrane and aqueous solution and α is the thickness of the membrane. Note that sodium and potassium are anions, but cloride is cathodic, therefore $[Cl]_i$ appears in the numerator in contrast to the anionic concentrations.

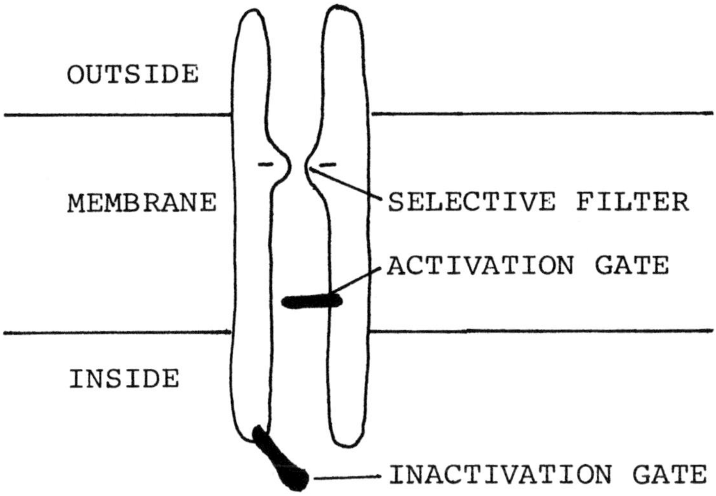

Fig. 3.2 *Sodium channel (simplified). The selective filter has a diameter of* 0.5*nm and allows only the small sodium ions to pass, but crossing the membrane is only possible if both the activation and the inactivation gates are open. The sodium channels react stochastically. In the resting state the squid axon closes nearly all the activation gates, but about half of the inactivation gates are assumed to be open (Hodgkin and Huxley, 1952; see also next chapter).*

Ionic channels

The membrane is not a complete insulator for ionic transport. Different channels in the membrane allow specific (small) ions to pass. For example, when a Na-channel opens for a short time, sodium ions are driven by the high outside concentration into the cell. Even in the resting state nerve and muscle cells take up radioactive-marked sodium ions, e.g., the resting influx of sodium into giants axons of the cuttlefish *Sepia* is about $35 pmol/(cm^2.sec)$. On the other hand, this quantity is low enough that prepared squid axons work for hours even when they are activated to send thousands of spikes. In living cells, the ions lost via ionic channels are returned by ionic pumps, which utilize metabolic energy in order to overcome the electrochemical gradient.

BOX 3.2 **ELECTRIC NETWORK FOR MEMBRANES**

The electric properties of cell membranes are characterized by their resistances and capacities. With $1-2\mu F/cm^2$ the capacity of the membranes are relatively large because of the very thin sheets of the lipid bilayers. In contrast to their constant capacity, the resistance of cell membranes depends to very large extend on the voltage sensitive gating mechansim of the ionic channels. Only in cases close to the steady state can we approximate membrane resistances by constants of the order of $2000\Omega.cm^2$ (e.g. for lobster axons. HODGKIN and RUSHTON, 1946). But variance is found from less than $1000\Omega.cm^2$ up to $50k\Omega.cm^2$ in membranes with relatively few channels.

Now, we will calculate the voltage across the membrane needed to drive a current impulse of $1\mu A$ and $1msec$ duration through a patch of $1cm^2$ of a cell membrane with constant conductance [the number of open channels is assumed to be unchanged in spite of varying voltage]. We define V as the difference between the inside potential V_i and the external potential V_e.

At first we will disregard the capacity; the membrane is modeled by an ohmic resistance as shown in (A). As the voltage is proportional to the current

$$V = R \cdot I \tag{A}$$

we need a voltage square pulse of $V = 2k\Omega \cdot 1\mu A = 2mV$ to obtain the desired current pulse.

In case (B) only the influence of the capacity is regarded. The current flow is determined by the change of charge Q according to

$$I = I_C = \frac{dQ}{dt} = C \cdot \frac{dV}{dt} \tag{B}$$

For our example the slope of the voltage is constant within the impulse time and we get $\frac{dV}{dt} = \frac{1\mu A}{1\mu F} = 1mV/ms$.

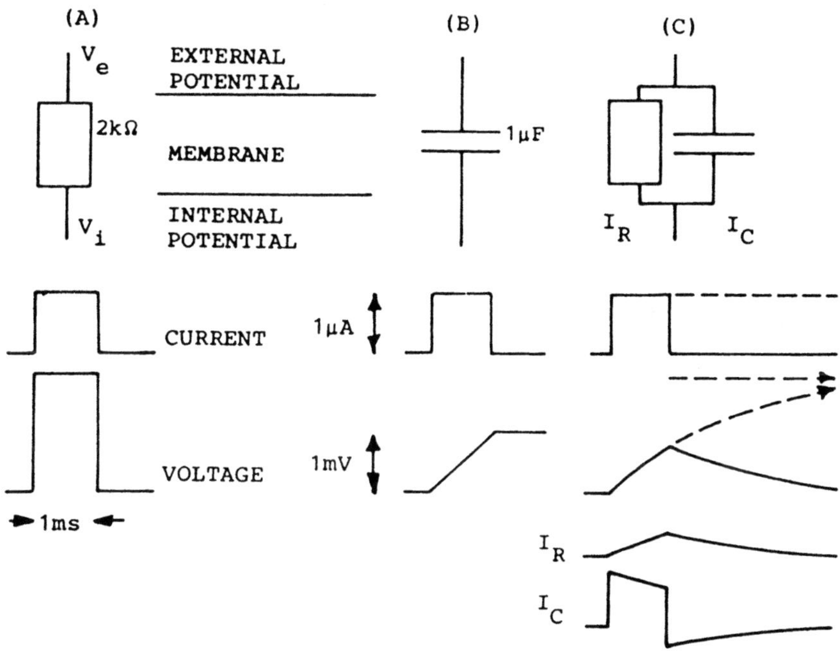

The voltage current relation for a network including both a resistance and a capacity is illustrated in (C). The total current I consists of an ohmic part and a capacity current

$$I = I_R + I_C = \frac{V}{R} + C \cdot \frac{dV}{dt}$$

and we will calculate V from the differential equation

$$\frac{dV}{dt} = -\frac{V}{RC} + \frac{I}{C} \qquad \text{(C)}$$

with $I = I_{max}$ in the interval $0 < t < t_{pulse}$ and $V_{t=0} = 0$. The result of this equation is

$$V = RI \cdot (1 - e^{-\frac{t}{RC}})$$

for $0 < t < t_{pulse}$.

The voltage exponentially follows the value which is defined by the pure ohmic resistance of case (A) (dashed lines in case (C)). The time constant of the membrane $\tau = R.C$ defines how quickly the transient behavior returns to the steady state. For our example we obtain $RC = 2000\Omega.cm^2 \cdot 1\mu F/cm^2 = 2msec$; note that RC has the dimension sec ! Using eqn. (C) we see that, at the end of the pulse, the voltage is defined by $e^{-\frac{t}{RC}} = e^{-0.5} = 0.6065$ which means that only 40% of the final value (of case (A)) is reached. ($V_{(t=1ms)} = 0.787mV$). After the pulse the voltage drops down to 0 with the same exponential behavior.

By splitting the total current across the membrane we calculate the ohmic part with $I_R = \frac{V}{R} = I(1 - e^{-\frac{t}{RC}})$ (within the pulse interval), and the capacitive part with $I_C = e^{-\frac{t}{RC}}$. See the lower traces of case (C).

The conductivity of a membrane, which is about $5nm$ thick, is extremely low compared with the fluid material on both sides. We obtain a specific resistance in the order of $10^{10} ohm.cm$ for the membrane; whereas, the resistivity of cellular fluid is about $100 ohm.cm$. This is a measure of the low permeability of cell membranes to small potassium, sodium and chloride ions which reduces the statistical chances of penetration by a factor of 10^{-8}.

The permeability of the membranes is a linear function of the times when the channels, embedded in the lipid layer, are open. The gating procedure depends on chemical parameters and on the voltage across the membrane. Several chemical substances have to be shown to be 'transmitters' which influence the gating mechanism in an exciting or inhibitive way. Acetylcholine (ACh), norepinephrine, epinephrine, gamma-aminobutyric acid (GABA) and a variety of other aminoacids, amines and peptides are these transmitters. Most of the synaptic excitation processes are initiated by transmitters, whereas, the gating procedure, needed for propagating signals in nerve and muscle fibers, is carried by the voltage of that part of the fiber which is already excited. A discussion on transmitters is beyond the scope of this book. For further information and references see, e.g.; KUFFLER *et al.*, 1984.

As an example, we will now consider the excitation process in a typical membrane which is mostly influenced by sodium and potassium ions. If enough Na-channels are opened by the influence of transmitters, a sodium influx increases the inside sodium concentration, resulting in a reduction of the voltage across the membrane. This causes more channels to be opened and an exponential rise of ionic transport starts. As a secondary reaction, ionic channels open for potassium ions and the K^+ ion efflux (flux from the inside to the outside) stops the increase of the inside potential and brings it back to the resting level.

Striking quantitative data on the voltage dependent conductances of sodium and potassium ions were gathered by the voltage clamp technique of HODGKIN & HUXLEY covered in the next chapter. Voltage clamp experiments supply us with statistical data on the gating processes involved whereas with modern techniques it is even possible to observe the currents passing through single channels of the membrane.

The patch clamp

NEHER, SAKMANN and their colleagues developed the 'patch clamp' technique to get single channel recordings. In 1976 NEHER and SAKMANN pressed a micropipette against the surface of a cell. Suction produced low pressure on the inside of the pipette to get a seal between the membrane and the glass of the pipette. In this way only those currents which pass through a very small patch of membrane were measured (Fig. 3.4). This technique was also used to cut a small patch of the membrane by a micropipette with an inside diameter of $0.3\mu m$ (Fig. 3.5). The patch, which contains only very few active channels is sealed against the pipette. Arbitrary solutions are put inside the pipette and in the bath before the current across the membrane is measured under voltage control. In this way, the patch clamp techniques were used to find both the open times and the resulting current strengths of single channels under the control of different intra- and extracellular solutions and also under the influence of transmitters.

In 1972 KATZ and MILEDI developed the 'noise analysis' technique on current noise produced by the fluctuation of open channels. On squid axons, single sodium channel conductance was

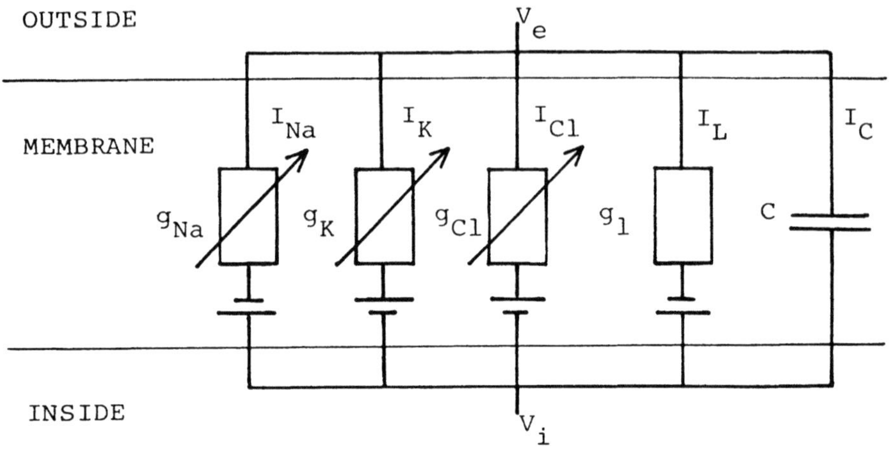

Fig. 3.3 Electrical model for a patch of cell membrane. In contrast to the model with constant conductances as described in Box 3.2, the membranes with active gating mechanisms have conductances (g_{Na}, g_K, g_{Cl} etc.) depending on the voltage $V = V_i - V_e$ across the membrane (V_i is the inside potential, V_e the extracellular potential). The additional voltage, resulting from the differences in the ionic concentrations on both sides of the membrane, is simulated by batteries drawn below the resistances. The sodium and the leakage battery want to produce an inward current, whereas, potassium and cloride tend to generate an outward current according to their electrical gradient which resulted from different concentrations on both sides of the membrane.

Each ionic conductance is directly related to the number of corresponding open channels. The total current is the sum of the ionic currents, a leakage current and a capacity current. In order to receive a quantitative description of the nonlinear ionic conductances as functions of voltage and time, Hodgkin and Huxley used the space clamp experiment. Furthermore, the knowledge of the quantitative influence of transmitters allows us to also use the electrical network to simulate activity of the membrane.

estimated by this technique with $4pS$ - the channel density with 330 channels per μm^2 (CONTIE *et al.*, 1975). At frog nodes of Ranvier, the sodium conductance was estimated with $6pS$ and channel density with $2000/\mu m^2$. In other tissues channel density ranges from $35/\mu m^2$ in *garfish* olfactory nerve up to $12000/\mu m^2$ in rabbit sciatic nerve (RITCHIE & ROGART, 1977).

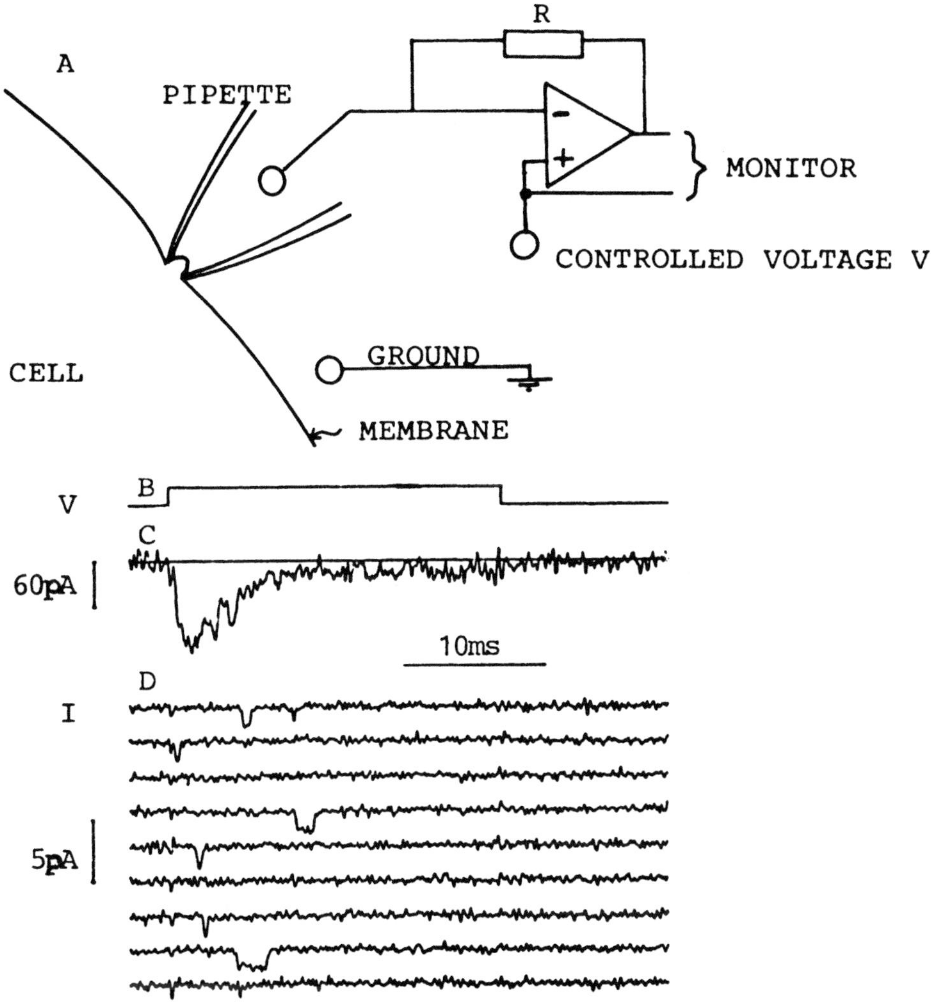

Fig. 3.4 *Single Na^+ channel current recording obtained by 'patch clamp' technique (whole cell recording). (A) A pipette filled with modified Ringer solution is pressed against the membrane of a cell. Suction (5-50 cm H_2O) is applied to the inside of the pipette to get a high sealing resistance. An operational amplifier records the currents passing through 2-5 Na^+ channels which are within the small area of the 0.3μm diameter opening of the pipette. 9 records of currents, produced by a 40 mV voltage pulse (B), are shown in (D). Superimposing 300 such records illustrates that most of the open channel times are found at the beginning of the voltage pulse (C). The mean single channel current was 1.6pA, the mean open time was 0.7ms.*

(After Sigworth and Neher, 1980)

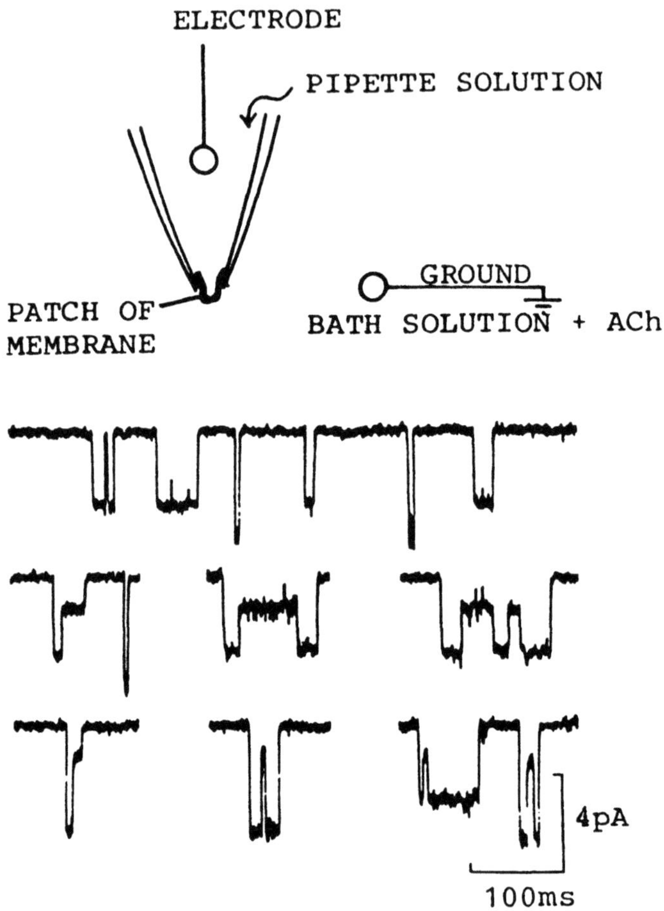

Fig. 3.5 *Single channel current from an isolated patch. Outside-out this means that the outside of the membrane faces to the outside of the bath solution. The gating mechanism is evoked by putting 2μmol/l ACh into the bath solution. At -100mV, 8°C the membrane of an embryonic rat muscle cell shows three different conductances leading to currents of 1.1pA, 2.9pA, and 4.3pA per channel, as illustrated by the recordings.*

(After Hamill and Sakmann, 1981)

Today most examinations are done on the voltage dependent sodium and potassium channels. But there also exists a lot of other important channels: We find ACh-channels at neuromotoric endplates. The conductance of Ca^{++}-channels depends on the concentration of Na^+ or K^+ ions, whereas, they have a good conductance to small univalent ions in the absence of bivalent ions, which may block the channel. The gating procedure of some channels depends on the concentration of intracellular substances. Most investigations in this field are done on the calcium dependent potassium channel, which reaches, at biological voltages, a good single channel conductance of $100 - 200pS$. This is at a high calcium concentration (some $\mu mol/l$); - but at the normal calcium concentration of a resting cell ($100nmol/l$), the channel opens only under a very high voltage. The combination of Ca-, K^+-, and calcium-dependent potassium channel is used by nature to produce rhythmic signaling, as well as, to control cellular processes like the contraction of muscle fibers or the secretion of hormones and neurotransmitters.

Chemical models show the different states which allow the ions to pass, but they are very complicated because there exist many different channels and every channel has several states. For our purpose, we will obtain quantative data for the complete excitation process by mathematical models, where statistical descriptions of the ionic conductances of the membrane will come from evaluations of suitable experiments. We are concerned with these models in the following chapters. At first we will consider the 'space clamp' experiment of HODGKIN & HUXLEY (1952a) to study the excitability of cells as a result of the nonlinear ionic conductances of the membrane. The electrical properties of the squid axon membrane can be modeled by the equations derived from these experiments, but most of the other membranes can also be simulated, eventually with some corrections, from the original equations. We will see in Chapter 6 that the propagating process of a firing axon can be simulated by an electrical network which obeys a system of HODGKIN-HUXLEY equations.

4. THE SPACE CLAMP EXPERIMENT OF HODGKIN AND HUXLEY – NON-PROPAGATING ACTION POTENTIALS

Measurement of the voltage-depending ionic membrane conductance

The cell membrane has pores which open up during excitation – this theory was postulated by JULIUS BERNSTEIN in 1902, but a quantitative description of the phenomena involved was found fifty years later by the ingenious experiments of HODGKIN and HUXLEY on giant squid axons. HODGKIN and HUXLEY assumed that a gating mechanism is responsible for the ionic transport across the membrane. Sodium and potassium ions are responsible for exciting the axons. They seemed to be independent from each other, but both can be described in a statistical manner. In order to quantify the voltage-dependent conductance of the membrane new electronic techniques were used.

Giant squid axons with diameters up to $1mm$ allow insertation of a stimulating electrode along the axis of the axon, (a second electrode was also inserted in order to measure the voltage between the inside and the outside of the axon.) In the 'space clamp' experiment there is no current flow along the axis and all parts of the membrane work under the same conditions because we get 'isopotentials' at the inside of the membrane as well as on the outside. We will call this a local model because it is often used to describe the reaction of the axon at a fixed point (without regarding propagation effects which cause variations of the membrane voltage along the axon).

The 'space clamp' experiment, according to Fig. 4.1, can be used to produce a simultaneous 'action potential' at all parts of the membrane by applying a current square pulse or a stimulus signal of arbitrary shape. But first of all we will consider the 'voltage clamp' experiment (the voltage is held constant) which allows to measure the ionic currents for controlled voltage steps. By exchanging the outside fluids or by using blockers which stop the activity of special types of channels, the ionic currents can

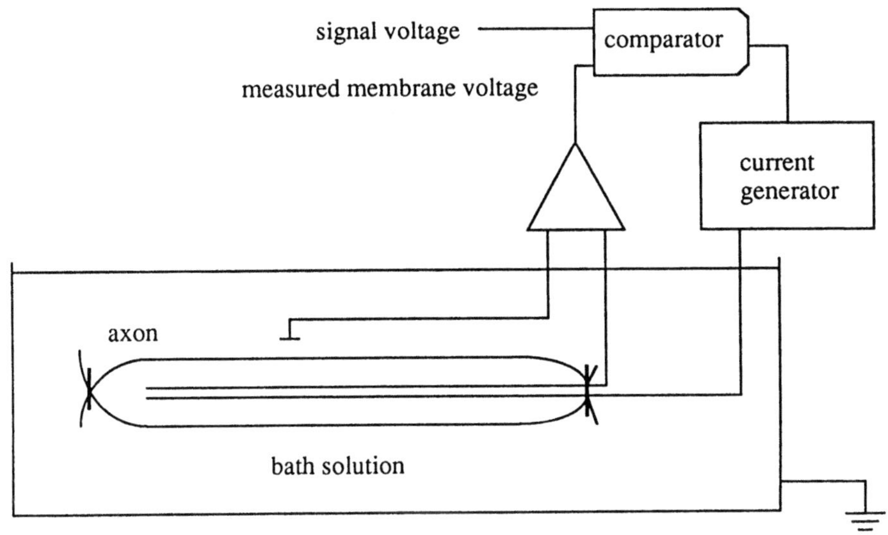

Fig. 4.1 Voltage clamp experiment according to Hodgkin and Huxley (simplified). Two long noninsulated wires are inserted along the axon which is sealed at both ends: One is used to record the voltage across the membrane, the other electrode injects just enough current to allow the recorded voltage to follow the signal voltage. This current is measured in order to analyze the nonlinear conducting behavior of the membrane.

be determined individually. Moreover, the axoplasm can also be pressed out and exchanged with other solutions in the perfusion technique. (BAKER *et al.*, 1961; OIKAWA *et al.*, 1961; ADELMAN & GILBERT, 1964; HUNEUS-COX, 1966)

All the current I_{inj} injected in the space clamp experiment has to pass the membrane as capacitive or ionic current. We can model the behavior of the membrane using the network of Fig. 3.3 together with eqn. (C) of Box 3.2

$$I_{inj} = I_R + I_C \tag{4.1}$$

or

$$I_{inj} = I_{ionic} + C \cdot \frac{dV}{dt} \tag{4.2}$$

In the voltage clamp experiment the time course of the voltage is given and the injected current is measured in order to find the nonlinear conductances for the ionic currents. For this purpose, the voltage is varied as a step or pulse function. In practice, this

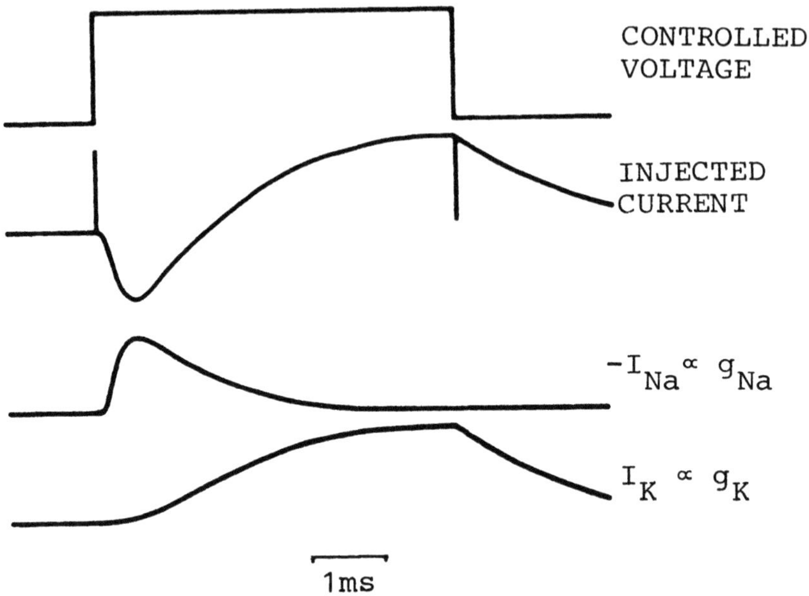

Fig. 4.2 *Voltage current relation observed in the voltage clamp experiment. In order to get the desired voltage pulse, the current generator (see Fig. 4.1) has to produce a signal according to the second trace. Besides the sharp current spikes, the whole injected current passes the membrane in the form of ionic current. The ionic current starts with a quick inward sodium current followed by a slow outward potassium current. As voltage is constant within the pulse time,* g_{Na} *and* g_K *are proportional to* $-I_{Na}$ *and* I_K*. Note that the sodium current of the Hodgkin-Huxley experiment is very similar in shape to the sodium current derived from statistical data of single-channel measurements in patch clamp experiments as shown in Fig. 3.4.*

Simplified after squid experiments of Hodgkin and Huxley, 1952a.

means that the voltage signal has a very large slope as the rise time is in the order of 1 μs. Within this short rise time interval a large current has to be injected and nearly all this current is needed to load the capacitor. After this short starting phase $\frac{dV}{dt} = 0$ demands to stop the capacity current and all the injected current passes the membrane in the form of ionic current until at the end of the voltage pulse a strong capacity current occurs again. [Within the pulse interval is $I_{inj} = I_{ionic}$.]

Fortunately, in the squid axon only sodium and potassium

ions are dominatly involved in the excitation process. The next task of HODGKIN and HUXLEY was to analyze the composition of the ionic currents. One way to separate them is to exchange the sodium component of the bath solution with larger cations which cannot pass through the small sodium channels; thereby the potassium outward current is found (lowest trace of Fig. 4.2). Because in the first experiment $I_{inj} = I_{Na} + I_K$, the sodium current is also determined.

The main result of the voltage clamp experiment is the time course of g_{Na} and g_K, which is proportional to I_{Na} and I_K. It is important to note that the electrical membrane behavior is not linear. This means that, e.g., doubling the amplitude of the voltage pulse does not produce just the doubled time course of I_{Na} and I_K. Therefore HODGKIN and HUXLEY varied the amplitudes of the voltage pulses and, by fitting the transient behavior of g_{Na} and g_K, they found for the squid membrane a description of the general current-voltage relation through four differential equations, which we will call the HODGKIN-HUXLEY (HH) MODEL. (see Box 4.1)

The HH equations are a very powerful tool for analyzing the membrane properties because they predict the membrane behavior for arbitrary shapes of stimulating signals. Although many other equations are in use, even today no other membrane model finds as many applications as the HH model. More about the method, further results and discussions, plus many mathematical investigations on the topic 'THE HODGKIN-HUXLEY AXON' are found in literature. Some interesting books and papers are: COLE, 1968; JACK *et al.*, 1983; SCOTT, 1977; CRONIN, 1981; MEVES, 1984; COHEN, 1976; and CONNOR *et al.*, 1977.

Quantifying membrane conductances

HODGKIN and HUXLEY assumed a model similar to the electric network shown in Fig. 3.3., where a constant leakage conductance g_L exists besides g_{Na} and g_K. Furthermore, they assumed that the sodium channel has barriers, according to Fig. 3.2, which can be described by a probability m for open activation gates and a probability h for open inactivation gates. In this way the sodium conductance (the coefficient g_{Na} in the HH equations below) will reach the maximum value only when $m = 1$ and $h = 1$ at the same time, i.e., all the sodium gates of the membrane are open! They assumed that potassium ions had to pass only a single type of gate which is controlled by the probability n. The time courses for these three probabilities are defined by three differential equations derived from the voltage clamp data.

In the following part the interested reader will find some details on finding the parameters to describe the gating mechanisms.

At first, a gating process is quantitatively represented by a single variable y being a function of time and voltage. y describes statistically the gating behavior of a high number of channels of one special type (e.g. potassium channel) lying in a small patch of membrane. $y = 1$ means all gates are open, $y = 0$ means all gates are closed.

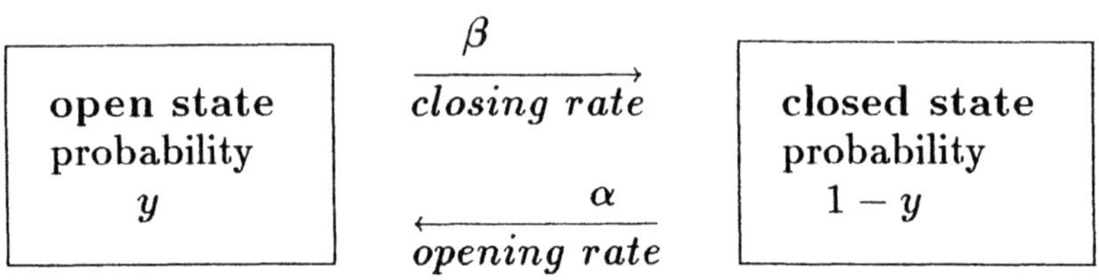

Fig. 4.3 *Simple gating procedure. At a fixed voltage, a constant β defines the change of the part of open gates which closes within a time unit, whereas at the same time $\alpha \cdot (1 - y)$ of the closed gates open up.*

The conductance of the membrane may be determined by the rate constants α and β. The probability $y(t)$ is the part of the open gates in the membrane. Within the time Δt the part $\beta \cdot y$ of open gates will close and from the part $(1 - y)$ of closed gates the quantity $(1 - y) \cdot \alpha$ becomes open. Thus, we can calculate the probability that gates are open with the help of a differential equation

$$\frac{dy}{dt} = \alpha(1 - y) - \beta y \tag{4.3}$$

Before a voltage step is applied the probability of open gates may be y_0 and according to equation (4.3) y increases exponentially to a steady state value y_∞ as shown in Fig. 4.4:

$$y = y_0 + (y_\infty - y_0)\left(1 - e^{-(\alpha+\beta)t}\right) \tag{4.4}$$

Table 4.1 Symbols, constants and [units] used in the Hodgkin-Huxley model

V	reduced membrane voltage (eqn 4.6)	$[mV]$
V_{rest}	resting potential	-70 $[mV]$
I_{inj}	current of electrode (eqn 4.7)	$[\mu A]$
i_{st}	stimulating current density	$[\mu A/cm^2]$
i_i	ionic current density	$[\mu A/cm^2]$
i_{Na}, i_K, i_L	sodium, potassium and leakage current densities	$[\mu A/cm^2]$
C	capacity of membrane	$[\mu F]$
c	capacity of membrane per cm^2	1 $[\mu F/cm^2]$
r	radius of fiber	0.024 $[cm]$
g_{Na}	max. Na-conductance	120 $[k\Omega^{-1}cm^{-2}]$
g_K	max. K-conductance	36 $[k\Omega^{-1}cm^{-2}]$
g_L	max. leakage-conductance	0.3 $[k\Omega^{-1}cm^{-2}]$
V_{Na}	voltage generated by different sodium concentration on both sides of the membrane	115 $[mV]$
V_K	potassium voltage	-12 $[mV]$
V_L	leakage voltage	10.6 $[mV]$
m, n, h	probabilities for opening the ionic channels	
k	thermic coefficient	
T	temperature	6.3 $[^oC]$
t	time in	$[msec]$

Considering the function $y = e^{-t/\tau}$, we see that after the time $t = \tau$ the value of y decays to $1/e$. As τ is the time scaling parameter for exponentially changing processes, τ is called the time constant. In our case of equations (4.3), (4.4) the time constant is

$$\tau = \frac{1}{\alpha+\beta} \qquad \text{and} \qquad y_\infty = \alpha \cdot \tau \tag{4.5}$$

BOX 4.1 THE HODGKIN HUXLEY EQUATIONS

The HODGKIN HUXLEY EQUATIONS describe the voltage–current relation for the membrane of the squid axon. It is convenient to use the reduced voltage

$$V = V_i - V_e - V_{rest} \tag{4.6}$$

where V_i and V_e are the intracellular and extracellular potential respectively and V_{rest} is the resting voltage of the cell, which is $-70mV$ for squid axons. Now, in the steady state $V = 0$.

Eqn. 4.2 gives a relation between the injected current and the voltage across the membrane, according to the voltage clamp experiment of Fig. 4.1, but these quantities depend on the diameter and the length of the fiber. In order to be independent of geometrical parameters, we will calculate the currents passing through $1cm^2$ of the membrane. Thus, all currents become current densities and c the capacity per cm^2. We obtain the stimulating current density i_{st} by dividing the electrode current through membrane area involved

$$i_{st} = \frac{I_{inj}}{2\pi r l} \tag{4.7}$$

where l is the length of the axon. Using i_{ionic} also as a current density, we obtain from (4.2)

$$\frac{dV}{dt} = [-i_{ionic} + i_{st}]/c \tag{4.8}$$

By setting $\frac{dV}{dt} = \dot{V}$ and $i_{ionic} = i_{Na} + i_K + i_L$, we arrive at the complete description of the HH model.

$$\dot{V} = [-g_{Na}m^3h(V - V_{Na}) - g_K n^4(V - V_K) - g_L(V - V_L) + i_{st}]/c \tag{HH-1}$$

$$\dot{m} = [-(\alpha_m + \beta_m) \cdot m + \alpha_m] \cdot k \tag{HH-2}$$

$$\dot{n} = [-(\alpha_n + \beta_n) \cdot n + \alpha_n] \cdot k \tag{HH-3}$$

$$\dot{h} = [-(\alpha_h + \beta_h) \cdot h + \alpha_h] \cdot k \tag{HH-4}$$

with the coefficient k for temperature T (in ^oC)

$$k = 3^{0.1T-0.63} \tag{HH-5}$$

and α and β, which were used to fit the ionic conductances to the experimental data

$$\alpha_m = \frac{2.5 - 0.1V}{e^{2.5-0.1V} - 1} \qquad \beta_m = 4 \cdot e^{-\frac{V}{18}} \tag{HH-6}$$

$$\alpha_n = \frac{1 - 0.1V}{10.(e^{1-0.1V} - 1)} \qquad \beta_n = 0.125 \cdot e^{-\frac{V}{80}} \tag{HH-7}$$

$$\alpha_h = 0.07 \cdot e^{-\frac{V}{20}} \qquad \beta_h = \frac{1}{e^{3-0.1V} + 1} \tag{HH-8}$$

The resting state conditions

$$V(0) = 0, \quad m(0) = 0.05, \quad n(0) = 0.32, \quad h(0) = 0.6,$$

completes the HH model which was originally used with $T = 6.3^oC$ ($k = 1$) according to the experimental temperature. The values for the other HH-parameters are listed in Table 4.1.

By multiplying the main equation of HODGKIN & HUXLEY (HH-1) with c we get a current relation. The probabilities m, n and h determine the ionic currents. Even in the resting state there is a very small current flow. The sodium current $g_{Na}m^3h(V - V_{Na})$ depends in a nonlinear manner (eqn. HH-2 and HH-4) by a factor of m^3h on the activation and inactivation gates. Because m is only 0.05, the third power of m is very small. Thus we see, that in spite of the inactivity gates 60% open (h(0)=0.6), there is hardly any sodium current passing through the membrane.

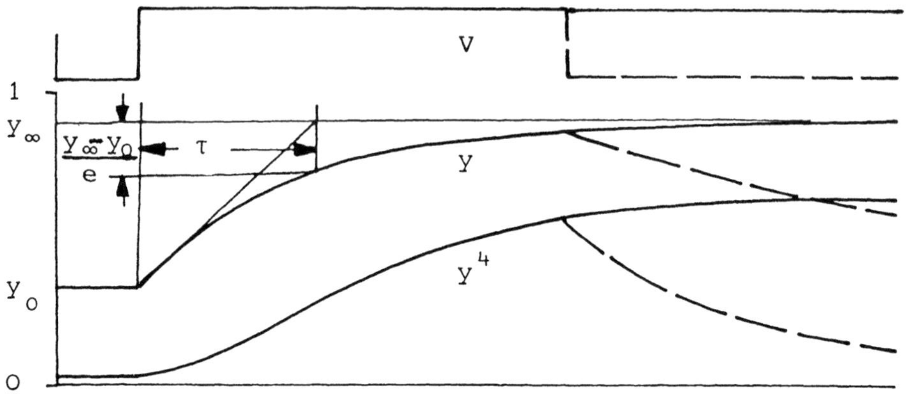

Fig. 4.4 *Reaction of gating mechanisms to an applied voltage step and to a 5ms voltage pulse (broken lines). The simple gating variable y changes exponentially from the starting value y_0 to the steady state y_∞. The falling edge of a voltages pulse stops the increase of y and causes exponential decay again to y_0. Note that the opening and closing rates, α and β, depend on the membrane voltage V. In the simple gating model, the probability y of open gates, is proportional to the conductance; whereas, in the advanced model, the conductance is proportionate to y^4, according to the potassium kinetics of the HH-model (compare Fig. 4.2).*

The computation was done for a voltage step from V=0 to V=60mV, with data of Box 4.1 (HH-7) for the potassium conductance, and therefore y corresponds with the HH-variable n:

$$\alpha = \frac{1 - 0.1V}{10.(e^{1-0.1V} - 1)} \qquad \beta = 0.125 \cdot e^{-\frac{V}{80}}$$

As V is the reduced membrane voltage, i.e., in the resting state V=0 and $y_0 = n(0) = 0.32$ (resting state condition, cf. Box 4.1) we find for V=0mV: $\alpha=0.058$, $\beta=0.125$, $\tau=\frac{1}{\alpha+\beta}=5.48ms$ and for V=60mV: $\alpha=0.504$, $\beta=0.059$, $\tau=1.78ms$ and for $y_\infty=\alpha\tau=0.895$. This means that after t=1ms, when the voltage jumps to 60mV, we must expect another time behavior compared with the end of the 5ms impulse. As $\tau=1.78ms$, y makes 63% (i.e.: $1-\frac{1}{e}$) of its way from y_0 to y_∞ within that time.

y_∞ and τ are the essential parameters of the gating process, giving information about the steady state and how quick it will be reached.

In living membranes the gating mechanism cannot be described by such a simple rule. Comparing, e.g, the time courses of Fig. 4.4 with the potassium conductivity as shown in Fig. 4.2 it is seen that the potassium conductivity starts very flat, but the slope of y of the simple gating model is $(y_\infty - y_0) \cdot (\alpha + \beta)$ (Fig. 4.4, eqn. 4.3). Using potences of y allows better fitting, and by approximation with y^4, good agreement with experimental data from squid membranes is possible. One could imagine that, in this case, four simple gates have to be passed one after the other to allow a ion to pass the membrane.

There is a remarkable difference between the conductances for potassium and sodium ions. The potassium conductance increases monotonicaly to the steady state, but in contrast, sodium conductance reaches a maximum and then again reduces to low values. HODGKIN & HUXLEY modeled this self-reducing process with the help of an inactivating gate. This gate is assumed to be at one end of the channel and in the resting state it has a high probability, h, to be open. Nevertheless, there is no good conductance for the Na^+ ions, as long as, the activating gates in the same channel have a low probability m to be open.

In the case of the giant squid axons, the ionic conductances can be well fitted using a combination of three simple gating mechanisms (as described by eqns. 4.3 and 4.4), quantified by m, n and h, respectively. Sodium conductance was fitted by $m^3 h$ and potassium conductance by n^4. In order to find the complete parameters for the HH-equations the following procedure was performed:

i) A sequence of voltage clamp experiments gives the time course of potassium– and sodium conductance for different voltage steps.

ii) Every experiment allows one to calculate a set of $\alpha_m, \beta_m, \alpha_h, \beta_h, \alpha_n, \beta_n$ for a fixed value of V.

iii) By data fitting all the α's and β's can be described as functions of V [see Box 4.1, HODGKIN & HUXLEY (1952)].

The influence of temperature

We have seen that membrane conductance depends essentially on voltage and time. Temperature also has an important influnce on the shape of AP's and the thermic dependence of membrane conductances is much more sensitive compared with other materials, e.g., with metals. [HODGKIN & HUXLEY (1952), HUXLEY (1959), FRANKENHAEUSER & MOORE (1963)]

The dependence of membrane conductance on temperature can be modeled by multiplying the right side of the equations for the gating variabales m, n and h with a constant k [eqn. (HH-2), (HH-3), (HH-4) of Box 4.1], making the reactions faster for higher temperatures. HODGKIN & HUXLEY have shown that all the gating processes of the squid axon reacts with about the same sensitivity to temperature steps. Usually, a special constant named Q_{10} is introduced, and Q_{10} gives the acceleration in membrane behavior when temperature is increased by 10°. HODGKIN & HUXLEY have found $Q_{10} = 3$ to be an adequate coefficient for squid axon membrane. They experimented at 6.3^oC, but the reaction at higher temperatures can be simulated successfully with $k = 3^{(T-6.3)/10}$.

Fig. 4.5 shows the influence of high temperatures on the form of the AP found from experiments and from simulation. Raising the temperature causes shortening of the AP's and a reduction of the maximum amplitude. The question is, whether AP's, with an essentially reduced strength, are able to propagate. When we use combined networks and standard HH-parameters to simulate the excitation process along the fiber*, it comes out that the AP's will not propagate for temperatures higher than 31^oC. A similar value of 33^oC was found by another method HUXLEY (1959). The 'heat block' was also observed experimentally by HODGKIN & KATZ (1949).

Stimulation with current impulses

We have seen above, that the voltage clamp experiments are useful to find a quantitative model for the excitation of a living cell. Now we will study the electrical stimulation of the space clamp cell by current injection. As mentioned above, an arrangement like that described in Fig. 4.1 allows us to inject a controlled current pulse. The result of such an experiment can be predicted by a simulation with the HH model.

---

* This will be described in detail in Chapter 6.

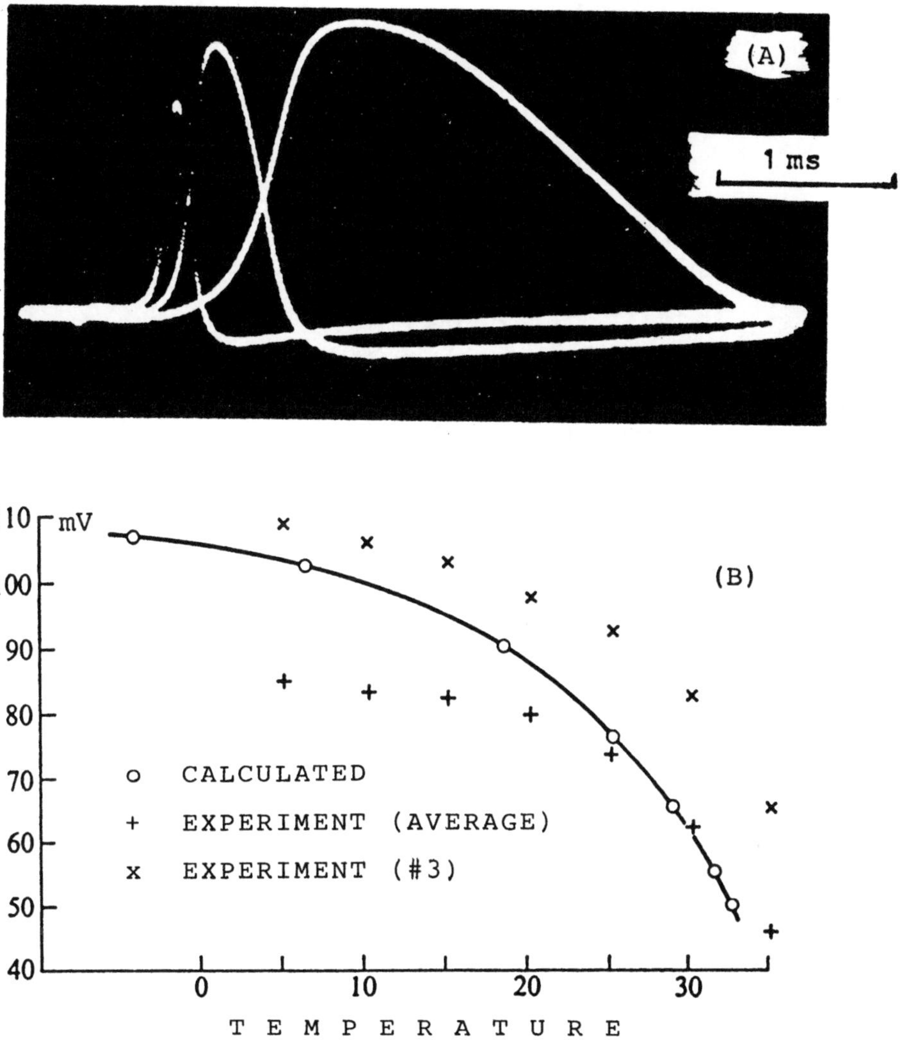

Fig. 4.5 Influence of temperature to the form of the action potential in squid axons. (A) Recorded AP's at 5°C, 18.5°C and 32.5°C show that pulse duration as well as spike amplitude becomes smaller when temperature is increased. (B) Spike amplitude as a function of temperature according to experiments from Hodgkin and Katz (1949) and computation with the HH-model.

(After Hodgkin and Katz, 1949 and Huxley, 1959)

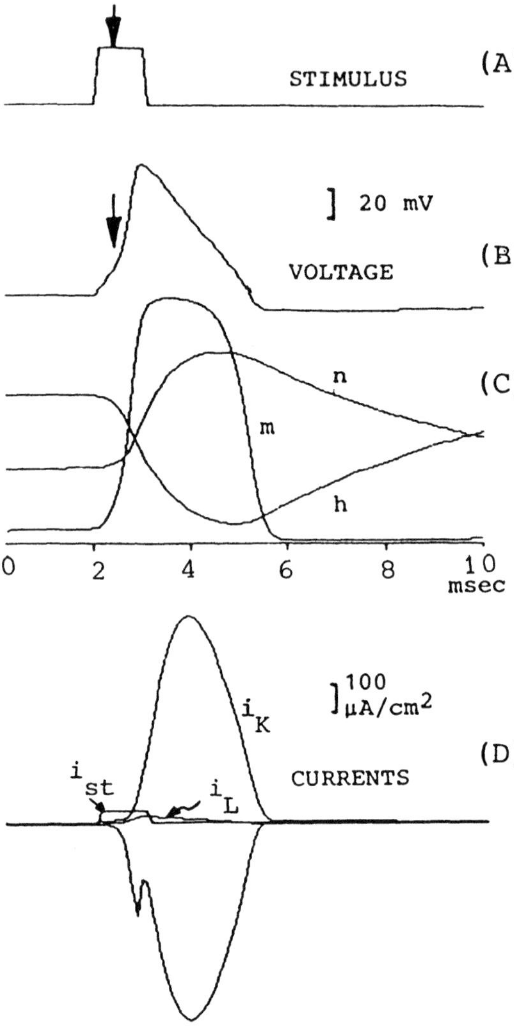

Fig. 4.6 Stimulation of a space clamped squid axon by current injection. A square impulse of electrode current stimulates the axon (A) which is at first in the resting state. Switching on the signal causes an approximate linear rise of voltage, which turns to an exponential rise when the threshold is reached (arrows) (B). The probabilities m, n and h are responsible for arising ionic activities (C). The action potential is started with a strong inward sodium current and then stopped by an outward potassium current which has a delayed start. In (D) also the square pulse of the stimulus current, which is much smaller than the ionic currents, is plotted. The leakage current is very small.

(After Rattay, 1987a)

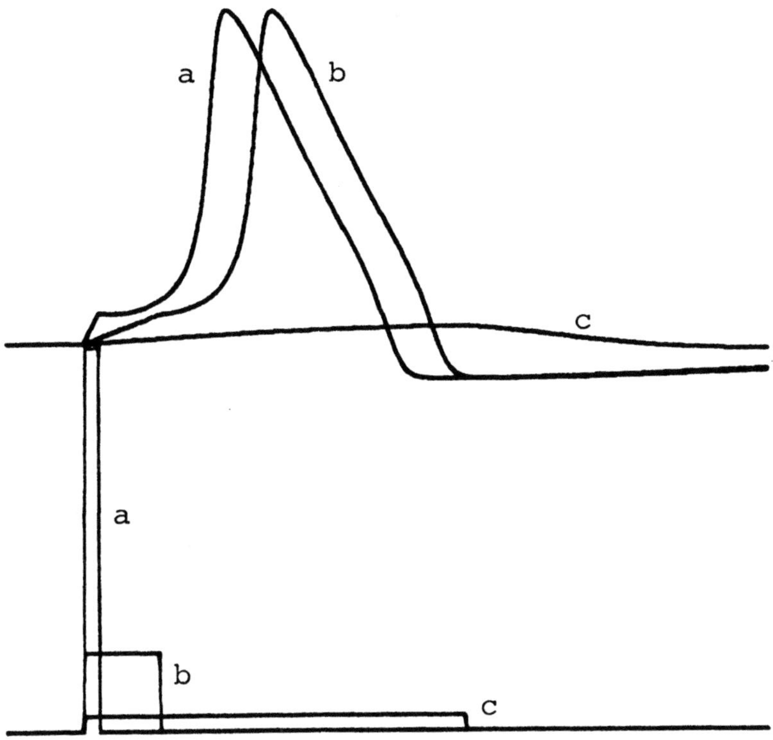

Fig. 4.7 Reaction to current impulses. Three impulses with the same charge but different duration are used to stimulate a squid axon. In the cases (a) and (b) threshold voltage is reached before the stimulating signal goes down. In case (c) the same charge shows only subthreshold response because some of the injected charge leaks away within the long stimulation interval.

Simulation with HH-standard data; current strength $100\mu A/cm^2$*, simulation time 10ms. Pulse duration: 0.2 ms (a), 1ms (b), and 5ms (c).*

Fig. 4.6 shows the response of the HH model when stimulated with a $1ms$ square impulse $i_{st} = 50\mu A/cm^2$. [From eqn. 4.7 we see that what we need for a squid axon with $r = 0.024cm$ (per cm of fiber length) is an electrode current of $I_{inj/cm} = i_{st} \cdot 2\pi r = 50 \cdot 2\pi \cdot 0.024 = 7.5\mu A/cm$ to obtain this result.]

The first quantity of injected charge is used to bring the membrane voltage to threshold (marked by arrows in Fig. 4.6). Fig. 4.6 (D) shows only small ionic flow before the time of firing. The sodium and potassium conductances depend on m^3h and n^4 and these quantities grow slowly from their resting values. The exponents 3 and 4 cause important influences of m and n only if they are greater than 0.5. Because of the small ionic conductances we get a rough estimation of the subthreshold behavior by reducing eqn. HH-1 to $\dot{V} = i_{st}/c$. This linear relation gives a reason for the straight subthreshold part of the voltage (Fig. 4.6 (B)). When threshold is reached, a rising sodium inward current becomes dominant and the voltage would finally reach V_{Na} as a stable state, provided no other ionic activities take place. Efflux of potassium ions bring the voltage back to the resting level. Sodium and potassium currents are much stronger than the stimulating current (Fig. 4.6 (D)); therefore, the voltage is not influenced considerably by that part of the stimulus current which is beyond the arrow (Fig. 4.6 (A)). Since the charges transported by sodium or potassium ions are about 50 times larger than that of the stimulating electrode, the axon is an amplifier of the stimulus signal. Note that, due to the notation of eqns 4.2 and 4.8, the stimulating current has the opposite sign as the ionic currents.

Fig. 4.7 illustrates the excitation caused by stimulating impulses with the same charge. Impulse duration of $1ms$ demands a current strength of $10\mu A/cm^2$ (case (b)). If pulse duration is shorter (case (a)) the same charge produces an earlier AP, but excitation fails for very long impulses (case (c)) because the ionic influences may not be neglected completely in the initial phase of stimulation. These nonlinear influences also cause small differences in the voltages at the end of the impulses in (a) and (b). Later, we will see that the "leakage effects" are greater in other local models, but it is especially apparent in spatial models, where stimulating current is applied at one point and this current can leak away *along* the fiber. [This is not possible in the space clamp experiment.] As a consequence of the leakage current it is impossible to reach threshold when current strength is below a certain value. This is called **rheobase.** We need, theoretically, an infinitely long impulse for $i_{st} = i_{rheobase}$ to obtain an AP. Another important measure is **chronaxie**, the duration of an impulse with a threshold amplitude of $i_{st} = 2 \cdot i_{rheobase}$ (Fig. 4.8).

Chronaxie is the classical measure of the responsiveness of the target neuron to temporal, as opposed to spatial, features of the electrical stimulation field. Chronaxies of myelinated fibers are more or less independent on diameter, and cluster around $100\mu s$.

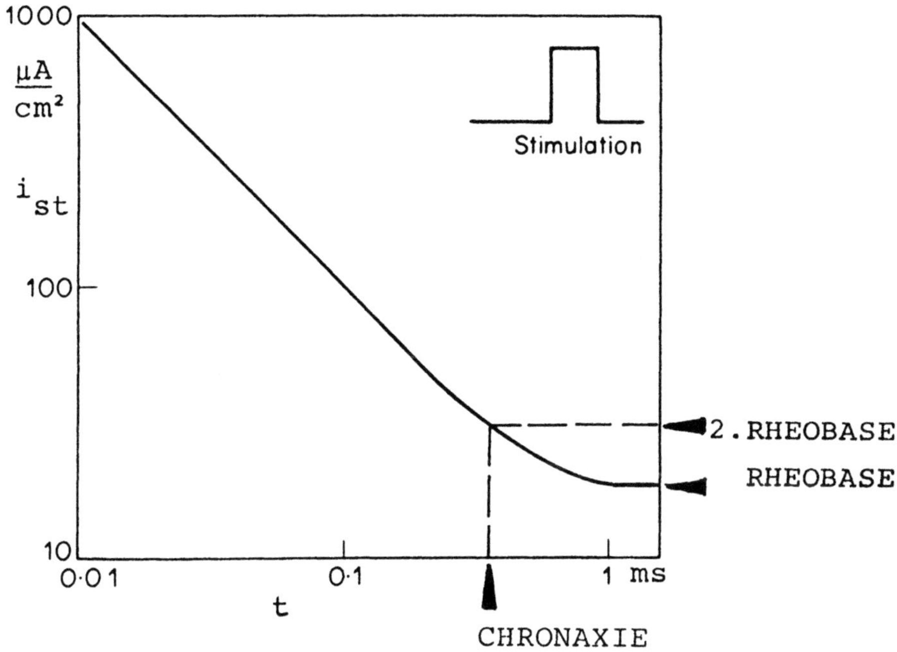

Fig. 4.8 Strength duration curve for positive injected current impulses calculated with HH standard data but for $T = 29°C$. For short times, the threshold stimulus charge is nearly constant. This causes the straight part in the strength duration curve which is plotted with logarithmic scales. Impulses with durations longer than 1ms need almost the same threshold current of $16\mu A/cm^2$, which is called rheobase. The dashed vertical line marks chronaxie, which is $340\mu s$ here. Chronaxie is the duration of that impulse which needs twice the rheobase current to generate an AP.

This value is in good agreement with model results and it is assumed that the time constant of the membrane is about 1.4 times chronaxie. The chronaxie of unmyelinated fibers and cell bodies is generally much larger (500-$600\mu s$), which is a consequence of the high capacity caused by the much larger amount of exposed membrane.

An interesting example, which will be discussed in detail in

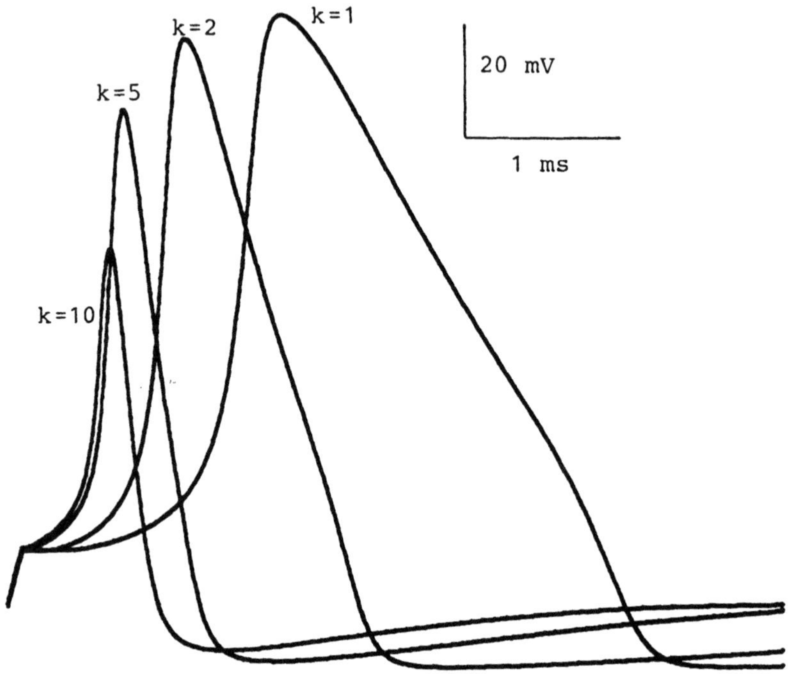

Fig. 4.9 *HH-response as a function of temperature. High temperature causes shorter refractory periods and smaller spikes. Coefficients $k=1$, $k=2$, $k=5$, and $k=10$ correspond with temperatures of* 6.3^oC, 12.6^oC, 21^oC *and* 27^oC *respectively. Stimulus current strength:* $100\mu A/cm^2$, *pulse duration:* $100\mu s$, *simulation time: 5ms. The simulated results are in accord with the original experiments of Hodgkin and Huxley. (See also Fig. 4.5)*

Chapter 12, is the excitation of the primary auditory nerve. We are especially interested in the fine timing structure of the firing pattern, which allows the user of a cochlear implant to discriminate between different signals thus making speech understanding possible. Surprisingly, the chronaxie of about $350\mu s$ as well as an AP-duration of about $300\mu s$ found experimentally is in good agreement with the results of the HH-model at 29^oC (RATTAY, 1986a; MOTZ & RATTAY, 1986; MOTZ & RATTAY, 1987; HARTMANN *et al.*, 1984; COLOMBO & PARKINS, 1987).

When we simulate the reactions of nerve fibers we should have in mind that special phenomena (blocking, anodal excitation) can occur as a consequence of the different spatial influences of the applied field. We will be concerned with this phenomena in Chapters 7–12. Nevertheless, within a certain range of the stimulus signal a first approach of the firing pattern can be found by simulation of the space–clamp experiment. As a simple example, Fig. 4.10 shows fiber reactions elicited by one or two $100\mu s$ pulses.

In the resting state, the inside potential is negative compared to the outside. One can expect that a positive (anodal) current will stimulate the fiber and a negative injected pulse will hyperpolarize the axon (Fig. 4.6 and Fig. 4.10). Stimulation from the outside should be done with the opposite sign, i.e., negative current pulses should be used, making the outside more negative. In Chapter 7, we will see that a more detailed analysis is necessary to understand the driving forces for the extracellular stimulation. It will also be proved why it is possible to produce an AP with the **inverse electrode current**, i.e., the question is why strong anodal extracellular stimuli are also able to produce AP's.

In contrast to all the other models the HH-model has a curiosity: it also allows the generation of an action potential by stimulation with inverse signals, that is with negative currents from the inside (Fig. 4.11). The mechanism for the **inverse inside excitation** is completely different from that for the inverse extracellular stimulation with anodal currents.

Normally, the AP will be driven by the Na^+ inward current which is plotted as a negative signal in Fig. 4.6. Thus, after the application of a stimulating impulse (I_{st} becomes 0 again) negative ionic currents are necessary in order to increase membrane voltage (HH-1). Fig. 4.11(B) shows the ionic currents across the membrane elicited by an $100\mu s$ impulse of $-400\mu A/cm^2$. When the stimulating impulse (which drives V to about $-15mV$) is switched off, all the ionic currents are negative, implying $\dot{V} > 0$. Note, that in the first phase V_L is dominant.

As a consequence of the relative slow gating variables marked for the case $I_{st} = -400\mu A/cm^2$ by m^-, h^-, n^- in Fig. 4.11(C), the voltage crosses the resting value $V = 0$. Now, I_{Na} becomes the dominant part of the total ionic currents and a normal AP is generated. In order to compare with the normal case Fig. 4.11(C) also shows the reactions of m, n, h for the AP evoked with $I_{st} = 100\mu A/cm^2$.

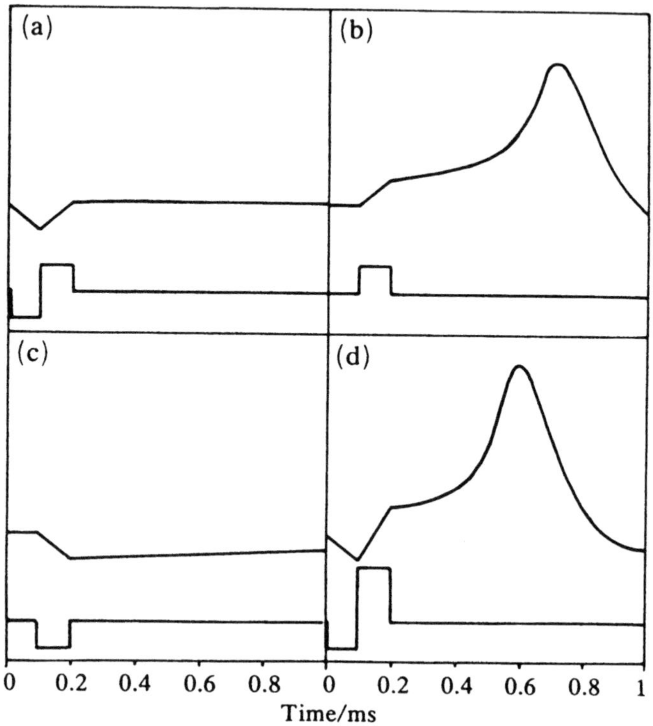

Fig. 4.10 *Fiber response (upper traces) to pulsatile electrical stimuli (lower traces). (a) A hyperpolarizing 100μs standard pulse is followed by a depolarizing one of equal duration and amplitude: the fiber remains at rest. (b) Starting from rest, a standard pulse (just above threshold) leads to an AP. (c) Once produced, hyperpolarization dies away very slowly. (d) A hyperpolarizing standard pulse followed by a depolarizing one of double amplitude produces an AP. Note that the stimulation effect in (d) is a little bit stronger than in (b). This is a consequence of the nonlinear membrane behavior.*

Simulation was done with the standard HH-data, but used a temperature of 29°C which caused a chronaxie of 350μs and an AP-duration typical for fibers of the acoustic nerve.

(From Motz and Rattay, 1987)

Summing up, it turns out that the special parameter configuration of the HH-model allows inside 'inverse' cathodic stimuli to produce AP's by a swinging through phenomenon similar to the hyperpolarization effect known from regular stimulation.

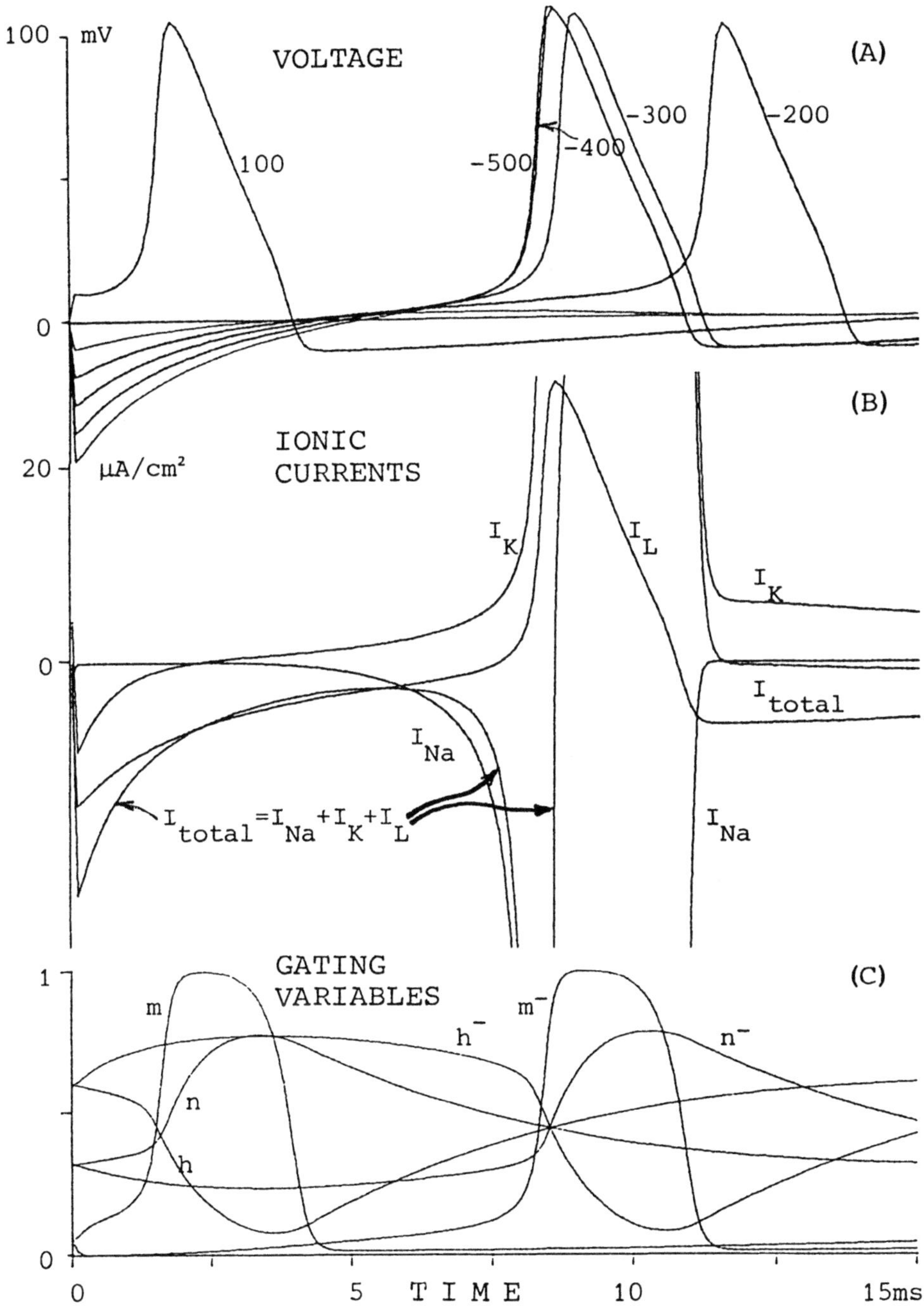

Fig. 4.11 The HH-model can also generate AP's by negative current impulses applied at the inside. Stimulation with standard HH-data; 100μs impulses with a strength of -500, -400, -300, -200 and +100 μA/cm² are used in (A). See text for details.

The nonlinearities of the HH-model make analytical treatment difficult and the pioneers had hard times with the calculators of the early fifties to find a single solution. The first extensive evaluations were done by COOLEY & DODGE in 1966. But today computer results for such a model are available very easily by using higher simulation languages. Nevertheless, we will also use simpler models, because they sometimes give more insight into the matter.

5. MODELING THE MEMBRANE

A reduced Hodgkin Huxley model

The number and types of ionic channels characterize the membranes. Sodium, potassium, chloride, calcium and calcium-dependant potassium channels are only some of the channel types involved in the excitation processes. The gating mechanism also depends on the concentration of transmitters and several other parameters. This situation seems to be confusing when trying to model the effects of electrostimulation. Fortunately, functional electrostimulation is concerned with the excitation of nerve and muscle fibers but not with synapses; therefore, we can exclude the difficulties coming from the transmitters and other chemical influences. For most of the applications, we can even neglect the differences in ionic concentrations resulting from high membrane activities. For the remaining differences in membrane behavior, which are found in fibers of interest, we find several models in literature. Besides the HH model we will be concerned in this chapter with the FRANKENHAEUSER-HUXLEY MODEL, which was introduced for myelinated axons, and finally we will study models for mammalian nerves which are interesting because only sodium and leakage currents are involved.

All these models have a similar behavior in reguards to the electrostimulation process and they can be described similar to the HH equations. Thus, all these models belong to the same class of differential equations: They have a stable steady state; small disturbances produce small excursions of the states, but higher influences bring them against a pseudo limit circle from where the trajectories come back to the resting level. We have seen this behavior at the HH model, but we will analyze the characteristics by reducing to two-dimensional models.

The HH model consists of four first order differential equations. We will reduce the four states V, m, n, and h to two variables. Fig. 4.6 shows that the states V and m have quite a similar shape and so do the slow variables n and h. Therefore, we can reduce the equations by taking advantage of the following relations: V is in the range from -10 to 115 mV and m varies from 0 to 1. The sum of the states n and h is nearly constant

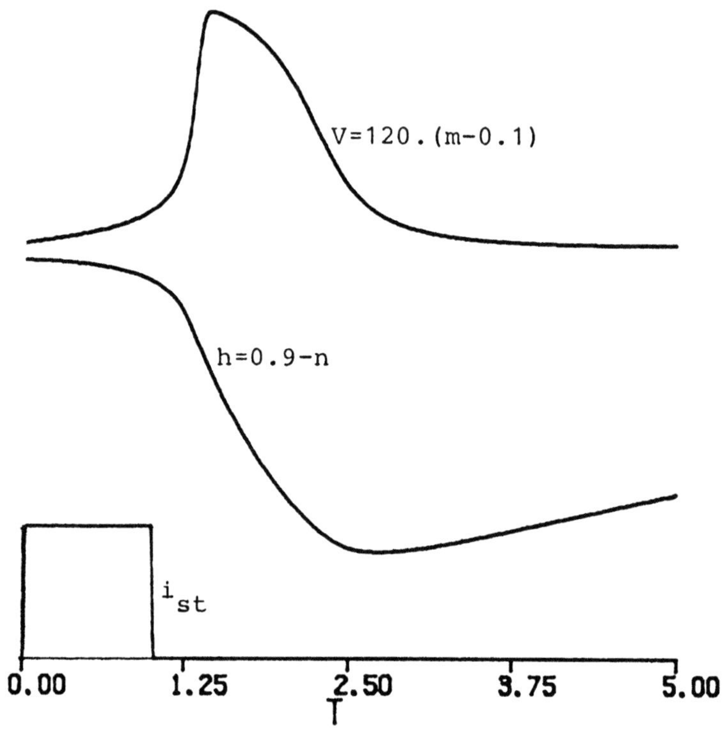

Fig. 5.1 Reduced Hodgkin-Huxley model. The HH model is reduced to the states V and h which are calculated from the equations HH-1 and HH-4 from Box 4.1 with HH standard data. Membrane voltage V has the same shape as m. Stimulus signal S is a 1ms square impulse.

with a value of about 0.9, thus we can introduce a reduced HH system, e.g., by both setting

$$m = \frac{V}{120} + 0.1, \qquad n = 0.9 - h$$

and by only using the equations HH-1 and HH-4. Comparing Fig. 5.1 with Fig. 4.6 we see that the form of the AP and the whole excitation process resemble each other closely.

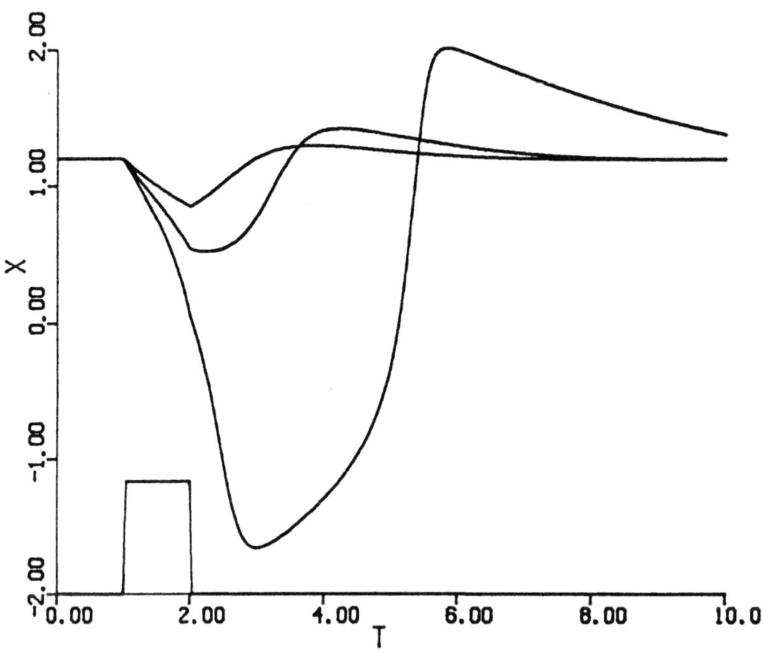

Fig. 5.2 *Reaction of the Fitzhugh model to square pulses with different current strength. 1ms current pulses with amplitudes s=0.5, s=0.75 and s=1 produce negative signals as nerve responses. Threshold: s=0.9.*

The Fitzhugh model

Another two-variable model, which is mathematically simpler, can be developed from VAN DER POL's differential equation. FITZHUGH uses an extension which is similar to BONHOEFFER's phase plane model; therefore the FITZHUGH MODEL is often found in literature as the BONHOEFFER VAN DER POL FITZHUGH MODEL (BVF). [BONHOEFFER 1943, 1948, 1953; FITZHUGH 1961, 1969; HOCHMAIR-DESOYER *et al.* 1984; RATTAY & MOTZ 1988]

The equations of the BVF-model are

$$\dot{x} = c \cdot (y + x - \frac{x^3}{3}) - s \tag{5.1}$$

$$\dot{y} = -(x - a + by)/c \tag{5.2}$$

where x may be interpreted as scaled voltage, s as the stimulus current density, and y as the recovery variable. All the results reported below are obtained with the standard FITZHUGH parameters

$$a = 0.7, \quad b = 0.8, \quad c = 3 \tag{5.3}$$

The AP form of the BVF model differs somewhat from that of the HH model (Fig. 5.2). Furthermore, voltage and time should be scaled to be comparable with experimental results.

We will now analyze the excitability of nerves on basis of the BVF-model with the help of the phase plane. In the resting state $\dot{x} = 0$ and $\dot{y} = 0$ hence we find from the BVF equations (5.1) and (5.2) that,

$$\dot{x} = 0 = y + x - \frac{x^3}{3} \tag{5.4}$$

$$\dot{y} = 0 = x - a + by \tag{5.5}$$

With the standard data for a, b, c (5.3) we get $x_{rest} = 1.2$ and $y_{rest} = -0.625$ as a stable solution. The behavior of the model in the phase plane is seen in Fig. 5.3 (B) and with more details in Fig. 5.4. The phase plane is a useful tool to analyze two first order differential equations. The courses of the trajectories only depend on the starting point and they go clockwise, as marked by arrows, to the resting point. At the highest and lowest point of the trajectories there is $\dot{y} = 0$ and all the extremity points are situated on the dashed line ($\dot{y} = 0$) of Fig. 5.4 which is defined by eqn. 5.5. The leftmost and the rightmost points of the curves are at the dashed line $\dot{x} = 0$, which is defined by (5.4) as the '$N - shaped$' curve $y = x^3/3 - x$. This N-shaped curve is typical for equations of nerve models. The N-shape of the curve $\dot{x} = 0$ (eqn. 5.4) allows the existence of a 'separatrix' which separates the small excursions from the long ways of the spikes. A solution, which starts at a point over the separatrix, simulates a subthreshold response. Most of the curves starting above the separatrix follow in short pathways to the resting state and it is difficult to find starting values for solutions which pass through 'no man's land'. [compare Fig. 5.3 (B)]. In practice, this means that we normally will not find a stimulus which can produce a substhreshold response that reaches, e.g., 90% of the amplitude of an AP.

The trajectories of Fig. 5.4 are solutions of the BVF model without a stimulus signal (s=0). Starting from the resting point, a stimulating signal has to drive the trajectory below (or, which

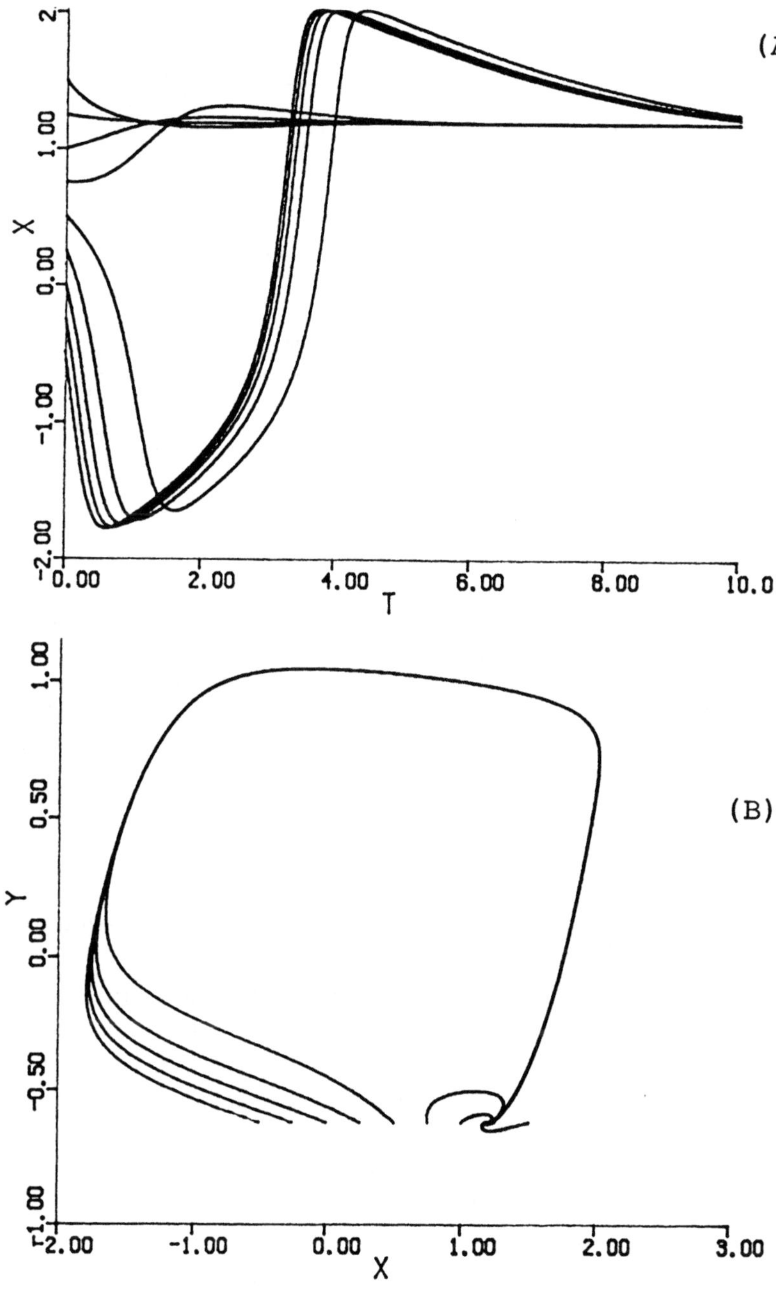

Fig. 5.3 *Response of the Fitzhugh model to different starting values x(0) without any stimulus signal. (A) Time course of the main variable x for x(0)= −0.5, −0.25, 0, ... 1.5;* $y(0) = y_{rest} = -0.625$. *(B) Phase plane diagram with the same starting values.*

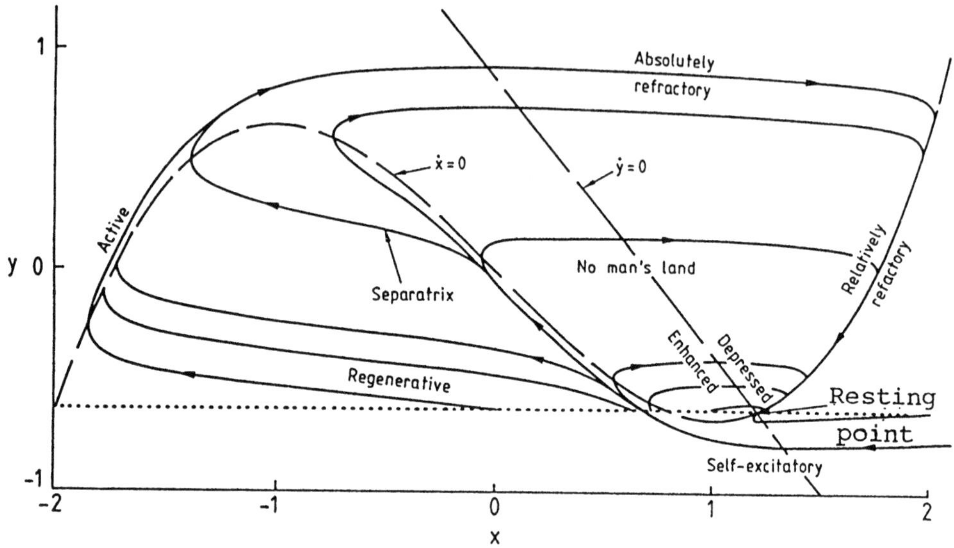

Fig. 5.4 Phase plane of the Fitzhugh model. All solutions of the Fitzhugh equations (without stimulus) finally end in the resting point. This is the intersection of the curves $\dot{x} = 0$ and $\dot{y} = 0$. Short pathways (subthreshold responses) are obtained if the initial values are above the separatrix. In order to get an AP the starting point must lie below the separatrix or a stimulus signal has to drive it there.

(After Fitzhugh, 1961)

is the same, to the left side of) the separatrix in order to produce an AP (Fig. 5.5).

Pairs or chains of pulses are of particular interest in functional electrostimulation. Inspection of the phase plane is also instructive in these cases. We have seen, from Fig. 4.6 and Fig. 5.2, that the after-potential reaction needs several ms to reach the resting state again. This time is called the **refractory period.** If we start a second stimulus signal within the refractory period we need higher threshold currents.

Fig. 5.6 illustrates this situation. The first impulse has an amplitude of 1 (threshold=0.9), whereas, the second impulse has the strength $s = 1.5$. The time between the pulses is $\Delta t = 2, 4, 6,$ and $8ms$. The stimulus signal is shown in Fig. 5.6 (A) for $\Delta t = 4ms$. As the reactions for different Δt are superimposed, the first part in all cases is the same. Only in the last case a further AP is produced [because the second impulse drives the trajectory

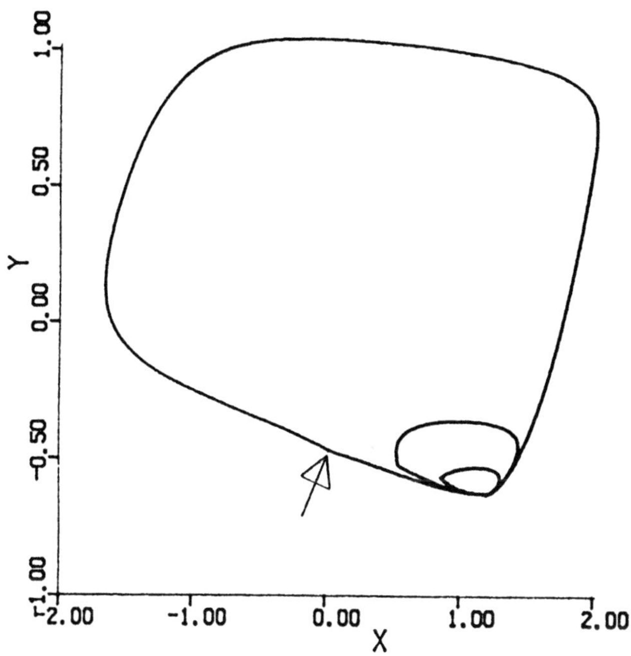

Fig. 5.5 Phase plane diagram for square pulse stimulation. Same data as in Fig. 5.2. Only the strongest of the three impulses shifts at the end (marked by arrow) the trajectory across the separatrix. Now (after 1 ms) there is no stimulus signal active and therefore the curve follows exactly the path which is determined by an initial value solution starting from the tip of the arrow, as described by the phase plane diagram of Fig. 5.4.

across the separatrix and the path in the phase plane is used for a second time. Fig. 5.6 (E)]. With $\Delta t = 6ms$, the continuing second stimulus becomes too weak and only a higher stimulus will help to reach the separatrix (Fig. 5.6 (D)). With $\Delta t = 4ms$ an even stronger stimulus is needed.

All these cases are within the **relative refractory period** (Fig. 5.4). If the second pulse starts in the **absolute refractory period** we can obtain only a single spike with a second peak. In the case of propagation within a fiber, such a double spike is soon reduced to a single AP, whereas, the small disturbance which was seen for $\Delta t = 4ms$ and $\Delta t = 6ms$ disappears.

In medical applications of electrostimulation, accumulation

BOX 5.1 THE FITZHUGH EQUATIONS

The FITZHUGH MODEL is simple to analyze and needs less computation time than experimentally fitted equations. Time transformation and scaling allow us to interpret the results physiologically. The modified equations give the nerve response to a stimulus signal $s(t)$.

$$\dot{x} = [c \cdot (y + x - \frac{x^3}{3}) - s] \cdot \beta \tag{BVF-1}$$

$$\dot{y} = \left[-\frac{x - a + by}{c} \right] \cdot \beta \tag{BVF-2}$$

For the standard parameters

$$a = 0.7, \quad b = 0.8, \quad c = 3 \tag{BVF-3}$$

the steady state values are $x_{rest} = 1.2$ and $y_{rest} = -0.625$. β is the time transformation factor, which changes the time scaling, but does not influence the form of the solution. Warm blooded axon excitation behavior can be modeled with $\beta = 4$ to $\beta = 7$ when time is measured in ms, i.e., the solution is 4-7 times quicker than for the original problem with $\beta = 1$. Membrane voltage V [in mV] can be approximated with

$$V = 25 \cdot (-x + 1.2) \tag{BVF-4}$$

These parameters have proved successful for several investigations (e.g. RATTAY & MOTZ, 1987), but other scaling will be used to fit to different experimental data.

of charge must be avoided because gassing would appear as a secondary electrolytical reaction. To avoid gassing a stimulating pulse may be followed therefore by a negative one of the same charge. As demonstrated in Fig. 5.7 it is possible that excitation

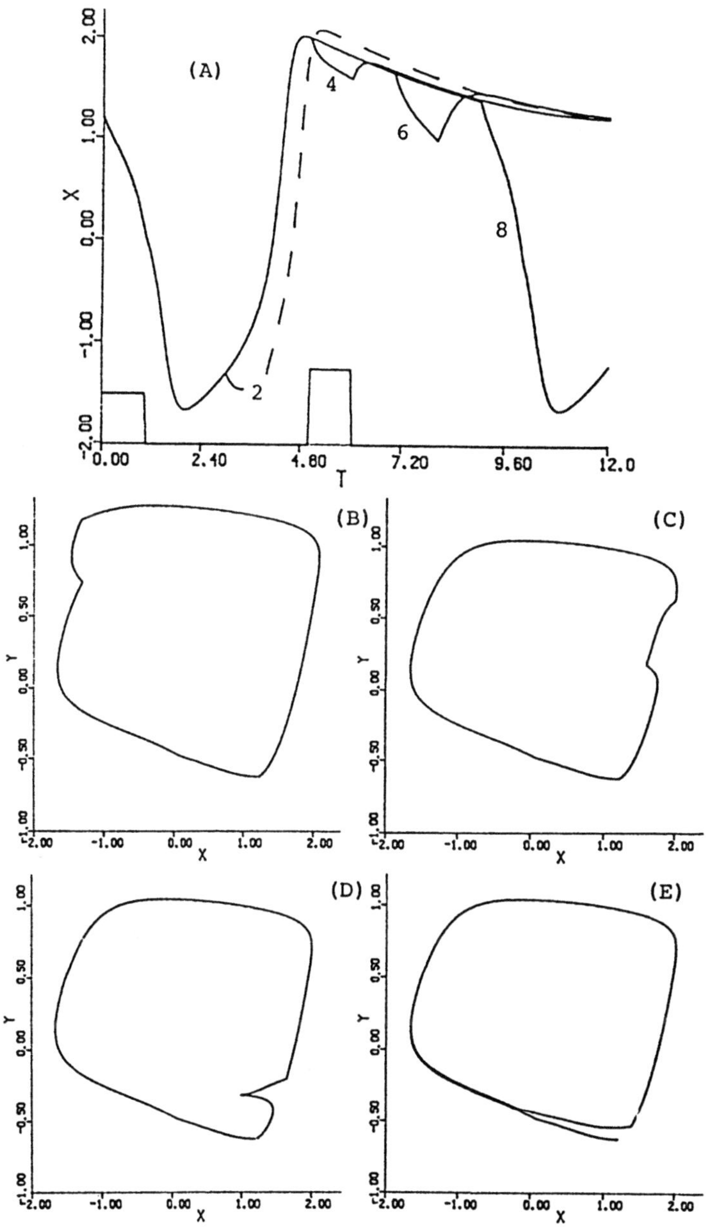

Fig. 5.6 Fitzhugh model response to a pair of positive impulses. A second (stronger) pulse follows after 2ms (dashed line), 4ms, 6ms, and 8ms, respectively; the first one of which produces an AP. All the solutions are superimposed in (A). A second spike is only produced when the second pulse starts 8 ms after the end of the first one. In the phase plane this corresponds with case (E), where a second circulation is started.

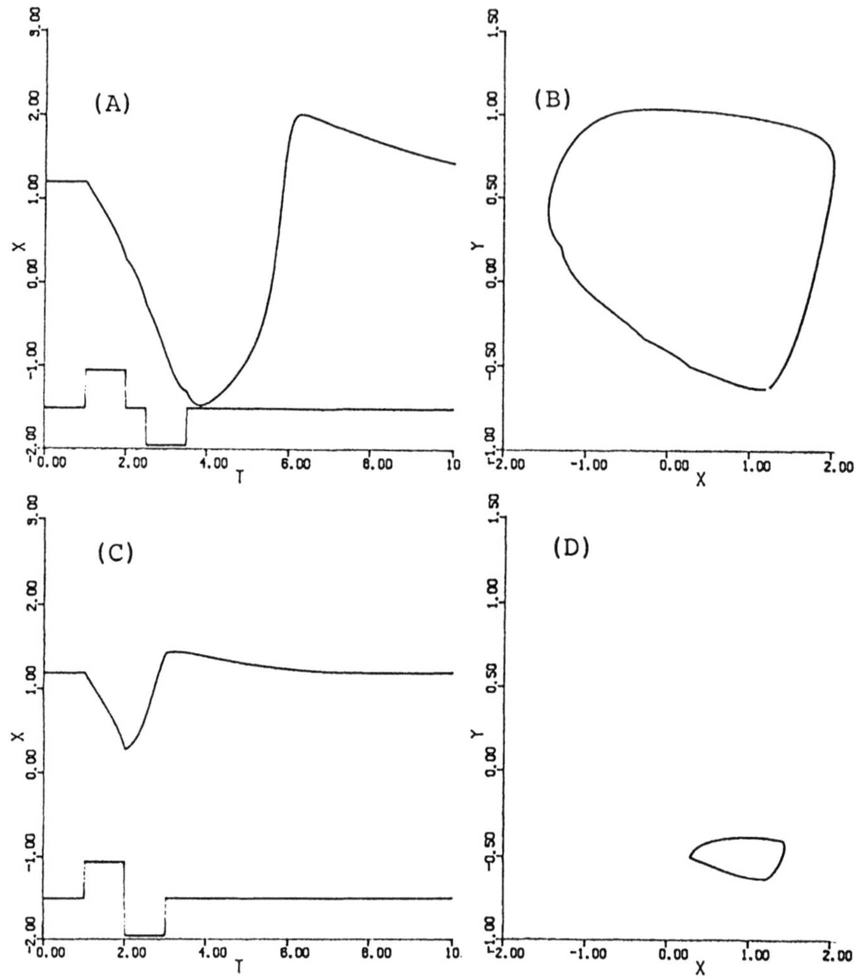

Fig. 5.7 *Influence of a second inhibitory pulse. A second negative impulse tries to bring the membrane voltage back to the resting level. However, if the excitation is continued, the second pulse will not stop the AP (A), because the negative signal is not strong enough to cross the separatrix (B). However, if the second pulse arrives sooner, it will stop excitation (C) and (D).*

fails for weak signals if the second pulse starts short after the first.

The Frankenhaeuser Huxley model

Myelinated axons are in a separate category of excitable cells. Myelin is a fatty layer covering most parts of the axon. Between

Table 5.1	Data of myelinated frog fiber
fiber diameter	14μ
thickness of myelin	2μ
distance between nodes	$2mm$
specific resistance of axoplasm	$110\Omega cm$
capacity per cm of myelin sheet	$10 - 16pF/cm$
dielectric constant of myelin	5-10
specific resistance of myelin	$500 - 800M\Omega cm$
capacity of nodal membrane	$3 - 7\mu F/cm^2$
resistance of resting node	$40 - 80M\Omega$
action potential	$116mV$
resting potential	$71mV$
peak inward current density	$20mA/cm^2$
conduction velocity	$23m/sec$

SCHWAN CELLS small gaps are free of myelin at the NODES OF RANVIER. Only in these free areas can ionic transport allow the excitation process. Table 5.1 gives some data of myelinated frog axons gathered by HUXLEY & STÄMPFLI (1949), TASAKI (1955), and HODGKIN (1964).

In 1964, FRANKENHAEUSER & HUXLEY summarized voltage clamp data for the frog node in the FRANKENHAEUSER-HUXLEY (FH) MODEL. It differs from the HH equations because they used permeabilties, after the GOLDMAN EQUATION, instead of using the conductances of the membrane. They also introduced a delayed nonspecific ionic current, I_P, mainly carried by sodium ions. With some discrepances, which are discussed below the myelinated FH model gives qualitatively the same responses as the HH model.

The FH-model was the most frequently used model to simulate the electrical stimulated fiber. Unfortunately, the main paper published in 1964 by FRANKENHAEUSER & HUXLEY, does not account for the essential thermal effects on the gating variables m, n, h and p. By neglecting the acceleration factor of the gating mechanisms, we miss the shortening of the AP's and the other essential feature changes caused by increased temperatures. The investigators who used the FH-model for simulations in the field of functional electrical stimulation had started at least 12 years after the publication of the model thus, most of them were not familiar with the original problems.

Table 5.2 Symbols, constants, and [units] used in the FRANKENHAEUSER-HUXLEY MODEL

V	reduced membrane voltage	$[mV]$
V_{rest}	resting potential	-70 $[mV]$
E	voltage across the membrane	$[mV]$
i_{st}	stimulating current density	$[\mu A/cm^2]$
i_i	ionic current density	$[\mu A/cm^2]$
i_{Na}, i_K, i_L, i_P	sodium, potassium, leakage and nonspecific current densities	$[\mu A/cm^2]$
c	capacity of membrane per cm^2	2 $[\mu F/cm^2]$
P_{Na}	sodium permeability constant	0.008 $[cm/s]$
P_K	potassium permeability	0.0012 $[cm/s]$
P_P	nonspecific permeability	0.00054 $[cm/s]$
g_L	max. leakage-conductance	30.3$[k\Omega^{-1} cm^{-2}]$
V_L	leakage voltage	0.026 $[mV]$
$[Na]_o$	external sodium concentration	114.5 $[mmole/l]$
$[Na]_i$	internal sodium concentration	13.7 $[mmole/l]$
$[K]_o$	ext. potassium concentration	2.5 $[mmole/l]$
$[K]_i$	int. potassium concentration	120 $[mmole/l]$
m, n, h, p	probabilities for opening the ionic channels	
F	Faraday constant	96485 $[C/mole]$
R	gas constant	8314.4$[mJ/K/mole$
T	absolute temperature	293.15 $[^oK]$ (=20^oC
t	time	$[msec]$
Q_{10}	thermic acceleration factor	
	for α_m	1.8
	β_m	1.7
	α_h	2.8
	β_h	2.9
	α_n	3.2
	β_n	2.8

For applications the main paper is misleading because there exist a temperature dependence in the equations (FH-2, FH-3, FH-4); however, this is of insignificant magnitude compared with that of the gating variables. For example, the values for

$$\frac{EF^2}{RT}\frac{[Na]_o - [Na]_i e^{EF/RT}}{1 - e^{EF/RT}} \quad \text{and} \quad \frac{EF^2}{RT}\frac{[K]_o - [K]_i e^{EF/RT}}{1 - e^{EF/RT}}$$

BOX 5.2 THE FRANKENHAEUSER HUXLEY EQUATIONS

The FRANKENHAEUSER-HUXLEY EQUATIONS describe the voltage-current relations for membranes in the nodes of Ranvier, i.e., at the sections of myelinated axons which are free of myelin. The structure of the model is similar to that of HODGKIN & HUXLEY (Box 4.1):

$$\dot{V} = [-i_{Na} - i_K - i_P - i_L + i_{st}]/c \qquad \text{(FH-1)}$$

with the ionic currents

$$i_{Na} = P_{Na} m^2 h \frac{EF^2}{RT} \frac{[Na]_o - [Na]_i e^{EF/RT}}{1 - e^{EF/RT}} \qquad \text{(FH-2)}$$

$$i_K = P_K n^2 \frac{EF^2}{RT} \frac{[K]_o - [K]_i e^{EF/RT}}{1 - e^{EF/RT}} \qquad \text{(FH-3)}$$

$$i_P = P_P p^2 \frac{EF^2}{RT} \frac{[Na]_o - [Na]_i e^{EF/RT}}{1 - e^{EF/RT}} \qquad \text{(FH-4)}$$

$$i_L = g_L (V - V_L) \qquad \text{(FH-5)}$$

where membrane voltage E is given by the reduced voltage V and the inside resting potential V_{rest}

$$E = V + V_{rest} \qquad \text{(FH-6)}$$

The gating variables are defined by

$$\dot{m} = -(\alpha_m + \beta_m) \cdot m + \alpha_m \qquad \text{(FH-7)}$$

$$\dot{n} = -(\alpha_n + \beta_n) \cdot n + \alpha_n \qquad \text{(FH-8)}$$

$$\dot{h} = -(\alpha_h + \beta_h) \cdot h + \alpha_h \qquad \text{(FH-9)}$$

$$\dot{p} = -(\alpha_p + \beta_p) \cdot p + \alpha_p \qquad \text{(FH-10)}$$

with the coefficients

$$\alpha_m = \frac{0.36 \cdot (V - 22)}{1 - e^{\frac{22-V}{3}}} \qquad \beta_m = \frac{0.4 \cdot (13 - V)}{1 - e^{\frac{V-13}{20}}} \qquad \text{(FH-11)}$$

$$\alpha_n = \frac{0.02 \cdot (V - 35)}{1 - e^{\frac{35-V}{10}}} \qquad \beta_n = \frac{0.05 \cdot (10 - V)}{1 - e^{\frac{V-10}{10}}} \qquad \text{(FH-12)}$$

$$\alpha_h = -\frac{0.1 \cdot (V + 10)}{1 - e^{\frac{V+10}{6}}} \qquad \beta_h = \frac{4.5}{1 + e^{\frac{45-V}{10}}} \qquad \text{(FH-13)}$$

$$\alpha_p = \frac{0.006 \cdot (V - 40)}{1 - e^{\frac{40-V}{10}}} \qquad \beta_p = -\frac{0.09 \cdot (V + 25)}{1 - e^{\frac{V+25}{20}}} \qquad \text{(FH-14)}$$

and the initial conditions

$$\begin{aligned} V(0) = 0, \quad m(0) &= 0.0005, \quad n(0) = 0.0268, \\ h(0) &= 0.8249, \quad p(0) = 0.0049 \end{aligned} \qquad \text{(FH-15)}$$

Note: At room temperature the factor RT/F is about 25 mV.

All the coefficients α and β have to be multiplied with

$$Q_{10}^{(T-293)/10}$$

for temperatures T measured in ^oK differing from the original experiment made at $293^oK = 20^oC$. The factors correcting the thermic behavior are of high importance, especially for applications concerned with the time structure of the AP.

Using the Q_{10}'s of Table 5.2, short AP's of only 0.3ms occur. It seems adequate to use smaller values, e.g., 1.4 for α_m and β_m and 1.8 for all the others.

Note: The thermic correction is not included in the original FH-model! One must be careful with published results on high temperature simulations, as up to now, this factor was generally neglected.

used in the eqns. FH-2, FH-3, FH-4, change when the experimental temperature ($20^0 C$) is raised to $37^0 C$. This change is for the resting voltage $E = -70mV$ 96% and 116% of their values, and 115% and 97% at maximum depolarization ($E = 40mV$), whereas there is no change for $E = 0$. Note, that these differences are in the order of the accuracy of some of the parameters of the model.

It may be that FRANKENHAEUSER & HUXLEY assumed that the reader is familiar with the thermic influences of the gating processes because FRANKENHAEUSER & MOORE published their findings on frog node thermic effects in 1963. They found that all the gating coefficients α and β have a Q_{10} of about 3 as in the HH model, but α_m and β_m have roughly half of this value. This means that the AP becomes faster for higher temperatures especially the falling phase.

In a rough approach*, we can simulate the myelinated fiber at temperature T according to FH by the original equations, but multiply eqn. FH-7 with $k_1 = 1.8^{T/10-29.3}$ and also multiply eqns. FH-8, FH-9, FH-10, with $k_2 = 3^{T/10-29.3}$. Due to the incorrect modeling, one must be careful with published high-temperature FH results. Also in this book, there are several results presented which are computed with the data published by FRANKENHAEUSER & HUXLEY in 1964. We will refer to such results by **original FH data**, but when using Q_{10}, we will refer to the results as **corrected FH data**.

Fortunately, most papers are concerned with the threshold behavior of single impulses. This is not very critical, as long as the axon is at rest before the impulse is applied, because the subthreshold reaction is mainly driven by the stimulating current. By inspection of the starting phase (Fig. 5.9) we see identical curves for all the four cases. If we take Q_{10} into consideration the exponential increase of I_{Na} (which ignites the AP) starts about 20% earlier (Fig. 5.9(B)). Thus, we can conclude that **starting at rest the corrected FH model needs roughly 20% less threshold current than the original model.**

The differences in thresholds between the original and the cor-

* As we know, most frog node parameters are accurate to only 1–2 digits. There is also a variance between the different fibers, hence one can expect only a rough quantitative estimation of reality by using model calculations.

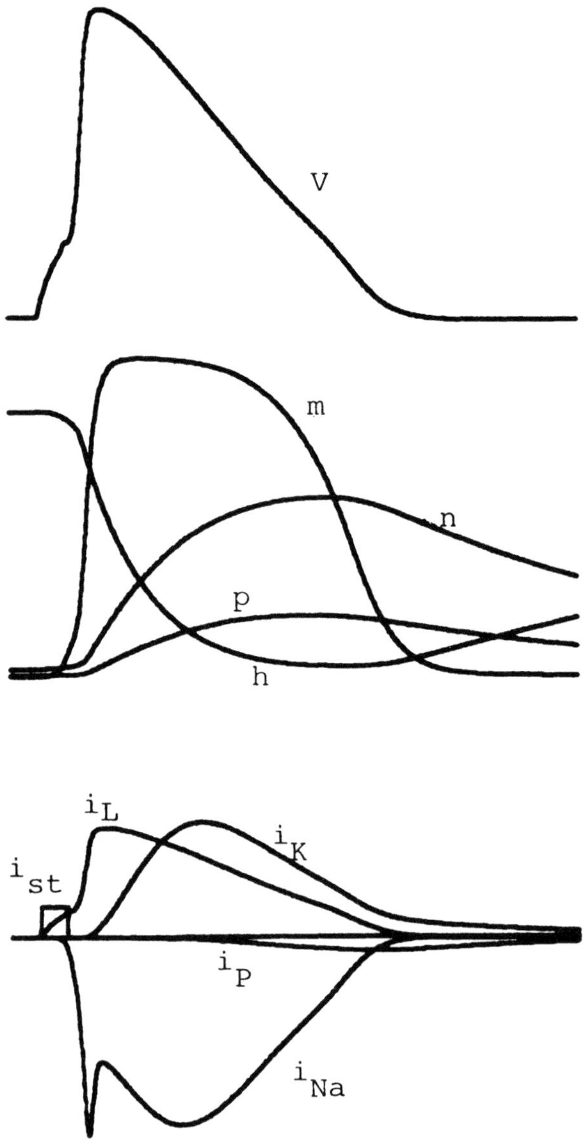

Fig. 5.8 States of the Frankenhaeuser-Huxley model. Voltage (top), gating variables (middle), and current densities at the membrane as consequence of the 0.1 ms stimulus impulse $i_{st} = 1mA/cm^2$ (bottom). As in all other figures, i_{st} has the opposite sign as the ionic currents.

Calculated for 2 ms with original FH standard data.

rected model are little higher when biphasic impulses are applied; this effect becomes more dominant for short pulses (Chapter 9). More striking however is the different behavior of both FH versions concerning the cathocic block. Strong cathodic extracellular stimuli will block excitation. This phenomenon is evident from experimental data, and is also seen in the corrected model. However, it is absent in the original FH model (Chapter 8).

FRANKENHAEUSER & MOORE calculated Q_{10} values from three different experiments made at $T = 2.5^O C$, at a mean value, and at $T = 20^oC$. Using these Q_{10}'s, we obtain extremly short AP's of about $0.3ms$. If we assume a duration of $0.4ms$ to $0.5ms$ for α-motorneurons, a proper AP duration can be obtained by setting $Q_{10} = 1.7$ for all α's and β's (BÜTIKOFER & LAWRENCE, 1978). Smaller Q_{10}'s will reduce the differences in the results of the original and the corrected FH model.

Comparison of the Hodgkin-Huxley and the Frankenhaeuser-Huxley models

We will compare now the answers of the FH and HH model when stimulated with a square pulse stimulus. (Fig. 5.8 and Fig. 4.6). For both models the AP shape is roughly similar, and only small deviations occur in m, n and h. For the FH model, the 'probability' states come closer to 0, but they appear with lower exponents in the equations, too. New is a late ionic current i_P of minor amount. i_P reaches its maximum at the end of the AP. Here its magnitude is of the order of i_K; therefore, i_P will only influence the late falling phase and the afteroscilation of the AP.

There are three striking differences in the models.

i) In contrast to the HH-model the leakage current is much larger and in the order of the potassium current.

ii) The FH-model is much more sensitive in the threshold region (Fig. 5.10): If the stimulus is just below threshold, the frog myelinated axon drops rapidly down, whereas, the HH-model has a flat slope for subthreshold state, as well as, for the stimuli just over the threshold, which produces a considerable delay (Fig. 5.10 (B)). Even in the superthreshold case, the FH spike goes a little down when the stimulus signal is switched off (Fig. 5.10 (A)).

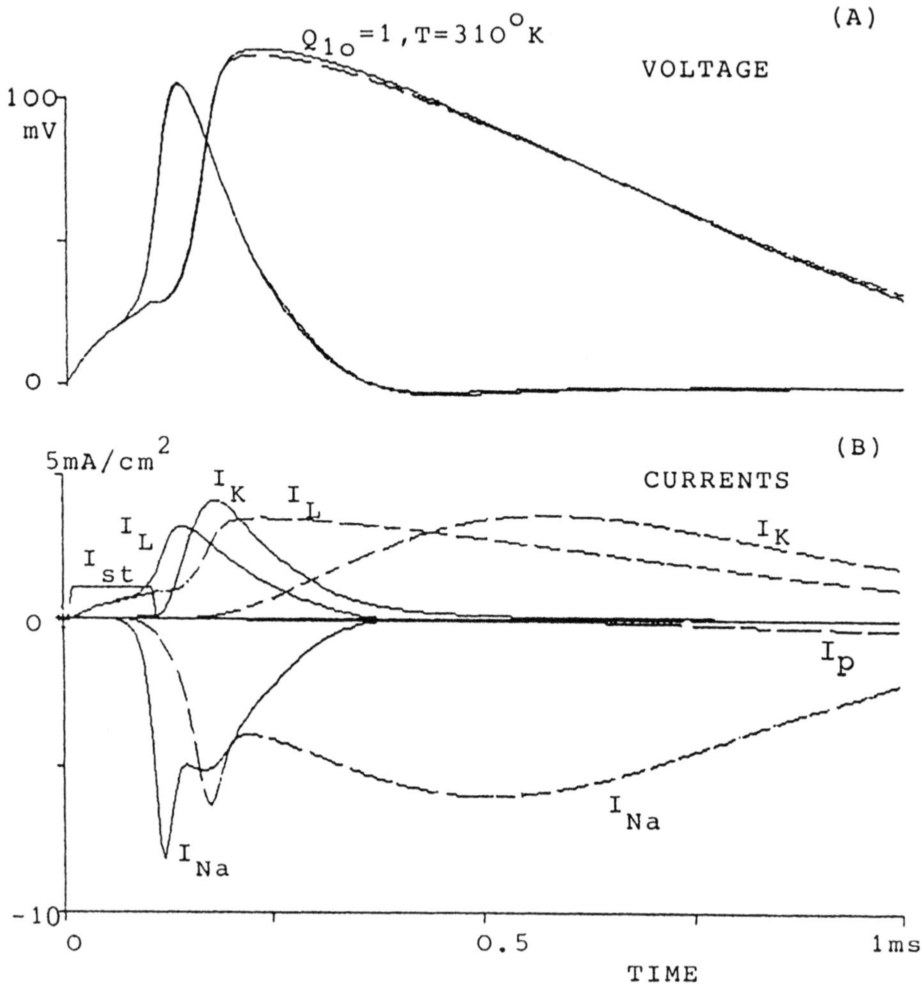

Fig. 5.9 Change of the FH action potential with temperature. (A) four different sets of data are used to calculate the voltages. The original FH model, as used by nearly all investigators, shows very similar <u>*slow*</u> *results for T=20°C (broken lines) and for T=37°C. The* <u>*fast*</u> *AP's, which are also quite identical, are calculated with individual Q_{10}'s after Frankenhaeuser and Moore (Table 5.2) and with a simplified Q_{10}=1.8 for α_m and β_m and Q_{10}=3 for all other values (see text). (B) current densities for slow and fast models.*

Note, that the rising slopes of V are always similar, but the fast models have fast falling phases. In all cases, the AP's are essentially the same in the starting phase. The ionic currents become dominant when the threshold is reached. The termic correction results in lower threshold voltages. Stimulus: 100μs current pulse.

iii) The FH afterpotential shows only a very small hyperpolarization effect (the voltage does not become really negative).

Active membranes without potassium channels

We have seen above that, in contrast to the HH membrane, in the myelinated frog axon the leak current is as important as the potassium current. It seems that in mammalian axons, leakage current is even more effective. Also, just like the HH and FH-membranes, generation of the AP is a consequence of the activation of sodium gating; however, repolarization results from both the inactivation of the sodium channels and the strong leakage currents. In 1968 the first voltage clamp experiments in single mammalian myelinated nerve fibers were perfomed by HORÁCKOVA *et al.* in the rat. They found that these nodal membranes almost totally lack potassium currents.

RICHIE and his colleges gathered data with voltage clamp experiments in order to model mammalian myelinated fibers. They demonstrated the absence of potassium channel influence when modeling the activity at the nodes of myelinated fibers in the sciatic nerve of rabbits. (CHIU *et al.* 1979). Nevertheless, voltage sensitve potassium channels also exist in such fibers. Although they have no significant influence in the intact membrane, it is notable that two types of voltage-sensitive potassium channels (fast and slow) exist in the nodal, the paranodal, and in the internodal regions. However, no sodium channels seem to be present in the myelinated membrane. See also SHERRATT *et al.* (1980), KOKIS & WAXMAN (1987), and RÖPER & SCHWARZ (1989).

One of the consequences of extensive demyelination is blocking of conduction. However, mammalian axons, which have been chronically demyelinated by diphtheria toxin, can develop conduction through a demyelinated region. This implies that, after demyelination, sodium channels appear in the exposed region of the membrane (BOSTOCK & SEARS, 1978; BOSTOCK *et al.* 1981; CHIU & RITCHIE, 1981).

From their data, RITCHIE and his collegues obtained a model similar to that of HODGKIN & HUXLEY. Unfortunately, the authors did not publish all the data necessary for simulation, although the complete model was used in the paper to demonstrate

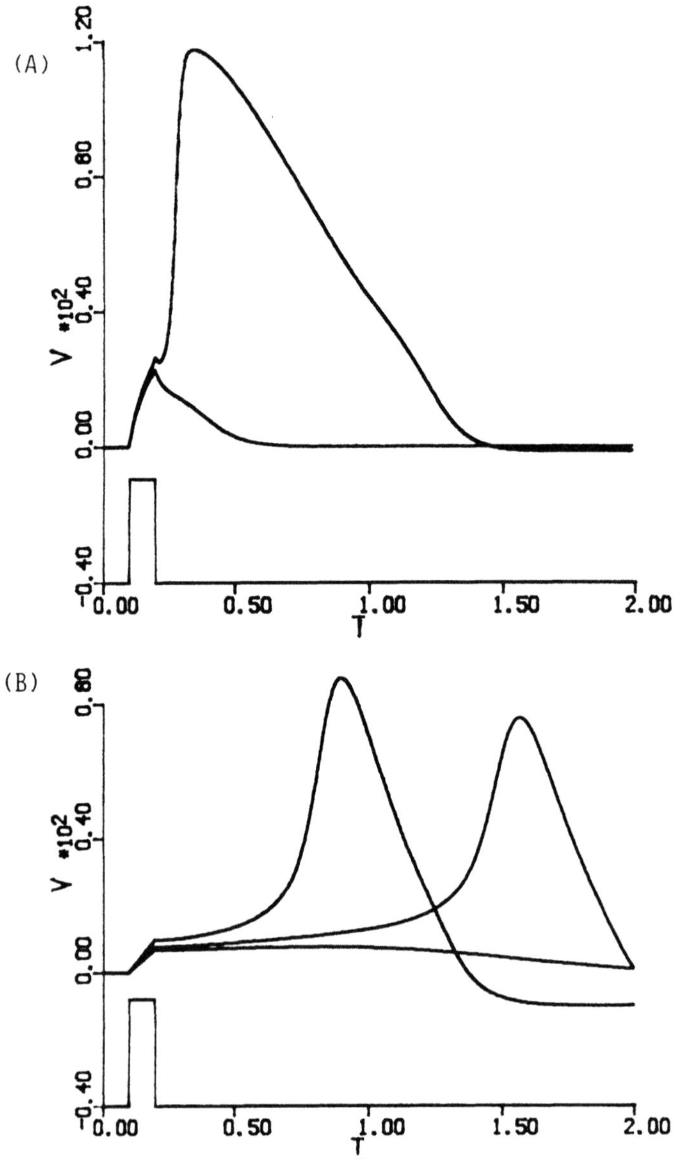

Fig. 5.10 *Comparison of the answers from the FH and HH models with stimuli near the threshold. Stimulation is with a* 100μs *pulse in all the cases. (A) FH responses to* $850 \mu A/cm^2$ *and to* $950 \mu A/cm^2$ *signal [*$37°C$*; original FH data, i.e.,* $Q_{10}=1$ *giving nearly the same results as for* $20°C$*] illustrates a breakdown at the end of the stimulus which arises quickly in the superthreshold case. The gating mechanism of the squid allows flat slopes in the surrounding of the threshold stimuli. This causes a considerabe time delay in the spikes (B). After the AP, the voltage drops remarkable far below the resting value.* $i_{st}=70$, *80, and 90* $\mu A/cm^2$, $T=21°C$ *in (B).*

BOX 5.3 A SODIUM AND LEAKAGE CURRENT MODEL FOR MEDICAL APPLICATIONS (CRRSS-MODEL)

In the myelinated rabbit nerves only sodium and leakage current seem to be important in the excitation process. SWEENEY *et al.* (1987) transformed the original data of CHIU, RITCHIE, ROGERT & STAGG (1979) to $T = 37^oC$ by applying an acceleration factor of

$$k = 3^{0.1T-1.4}$$

in accordance with (HH-5) of Box 4.1 and the experimental temperature of $T = 14^oC$. We name the following equations after the investigators the CRRSS model.

The experimental data of CHIU *et al.* (1979) was conducted by the method of DODGE & FRANKENHAEUSER (1958) after a modification of HILLE (1971). The model is similar to that of HODGKIN & HUXLEY without using the potassium current. The symbols are the same as those used in the HH-model (Table 4.1).

$$\dot{V} = [-g_{Na}m^2h(V - V_{Na}) - g_L(V - V_L) + i_{st}]/c \quad \text{(CRRSS-1)}$$

$$\dot{m} = [-(\alpha_m + \beta_m) \cdot m + \alpha_m] \quad \text{(CRRSS-2)}$$

$$\dot{h} = [-(\alpha_h + \beta_h) \cdot h + \alpha_h] \quad \text{(CRRSS-3)}$$

$$\alpha_m = \frac{97 + 0.363V}{1 + e^{\frac{-V+31}{5.3}}} \qquad \beta_m = \frac{\alpha_m}{1 + e^{\frac{V-23.8}{4.17}}} \quad \text{(CRRSS-4)}$$

$$\beta_h = \frac{15.6}{1 + e^{\frac{-V+24}{10}}} \qquad \alpha_h = \frac{\beta_h}{e^{\frac{V-5.5}{5}}} \quad \text{(CRRSS-5)}$$

The resting state conditions are

$$V(0) = -0.01mV, \quad m(0) = 0.003, \quad h(0) = 0.75 \quad \text{(CRRSS-6)}$$

The conductances are: g_{Na}=$1445mS/cm^2$ and g_L=$128mS/cm^2$; c=$2.5\mu F/cm^2$, $V_{Na} = 115mV$ and $V_L = -0.01mV$.

Note: The high slopes of the states of the model demand careful calculations.

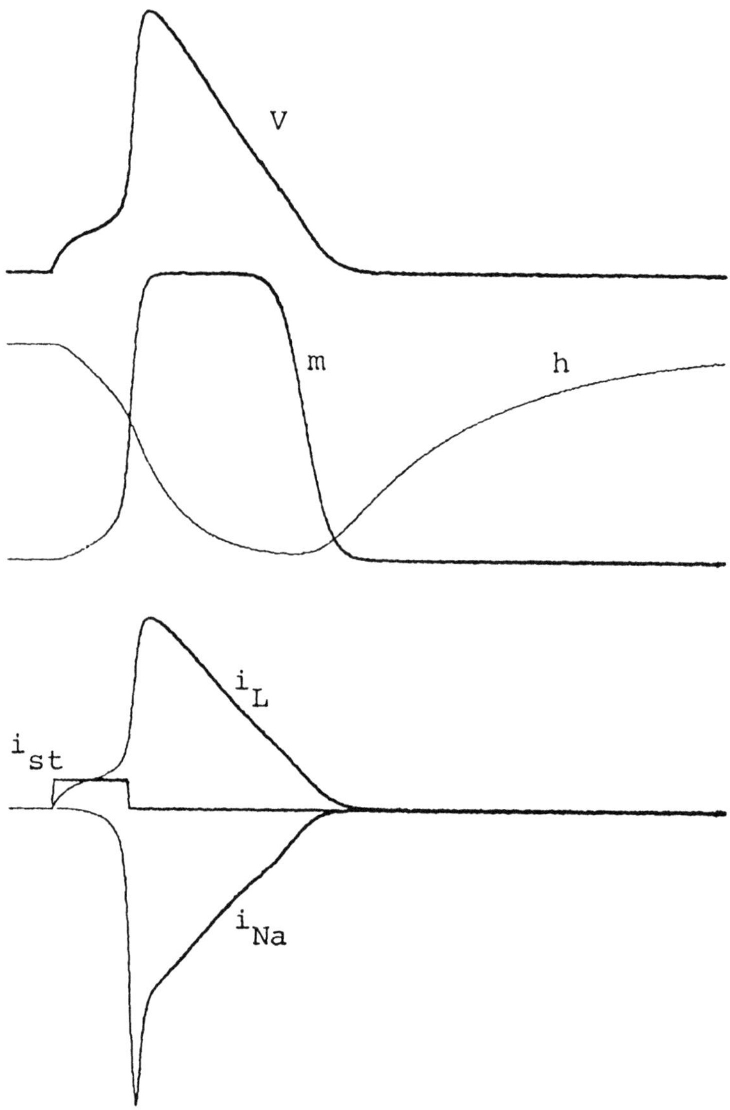

Fig. 5.11 *CRRSS – a model for mammalian myelinated nerve fibers. The membrane Voltage V has an interesting shape in the foot, which is followed by a sharp rising part, and very little hyperpolarization in the afterpotential. There are two gating variables, m and h, and two ionic currents involved. In contrast to the HH and FH model, only leakage current is needed for the recovery phase and the sodium current has only one peak.* $100\mu s$ *stimulation pulse of* $1800\mu A/cm^2$. *Within the simulation time of 1ms, all variables come close to the resting values. Calculation according to data of Box 5.3.*

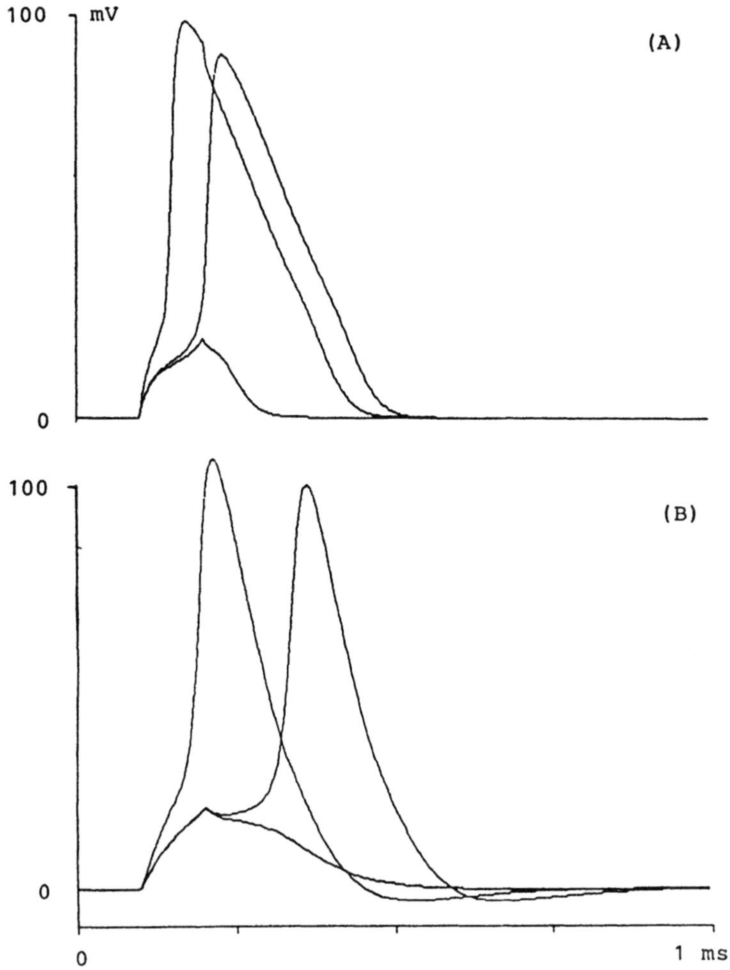

Fig. 5.12 *Answers of the CRRSS model (A) and of the FH model (B) to current impulses. In (A) three* $100\mu s$ *impulses of 1700, 1800, and 3000* $\mu A/cm^2$ *are applied (threshold* $1710\mu A/cm^2$*). In (B) the impulses have strengths of 720, 730, and 1600* $\mu A/cm^2$ *(threshold 728* $\mu A/cm^2$*). In contrast to the CRRSS model, FH has a hperpolarized tail. The AP's have similar durations, but using FH, the excitation needs longer to come to rest. This is consequence of the slow gating variables n and h.*

Simulation with standard data, but for the FH model Q_{10}*'s listed in Table 5.2 are used.*

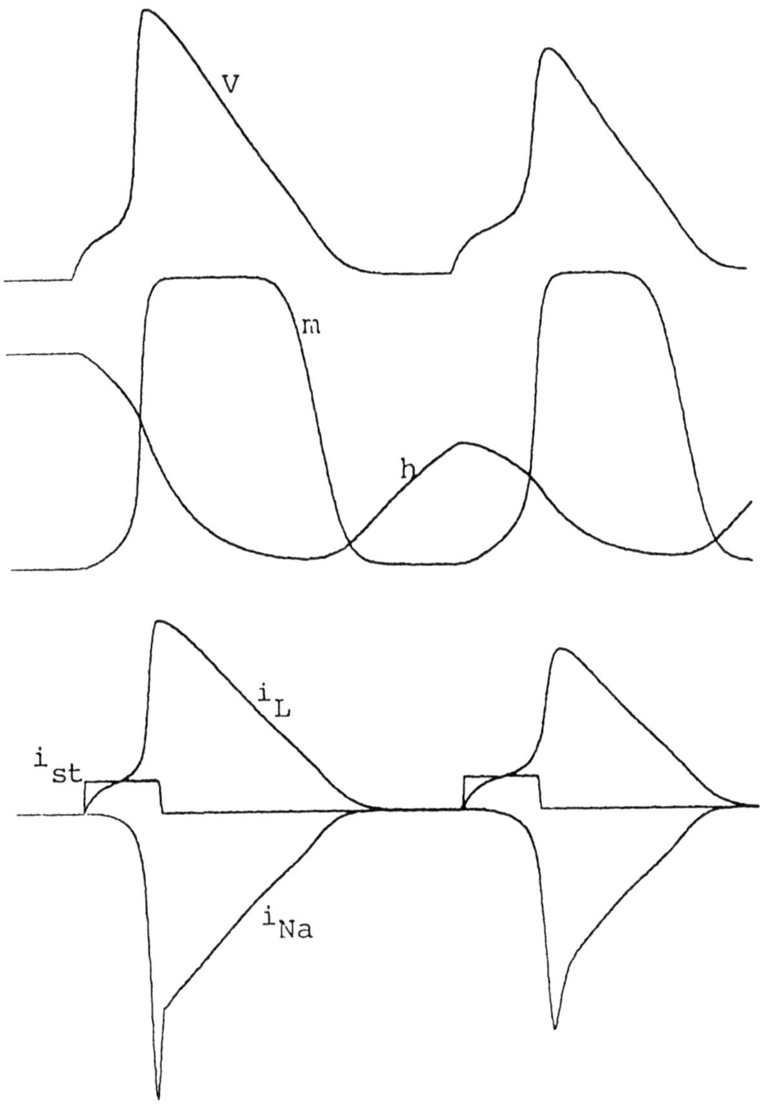

Fig. 5.13 *Refractory behavior of the CRRSS model. The gating variables m and h, and all the other variables react quickly. Therefore, it is possible to produce a second AP only 0.5ms after the first, even when weak signals are applied. Since h is below its resting value when the second AP is started, only a smaller voltage-peak is reached.*

Simulation with standard data; 100μs impulses of 2000$\mu A/cm^2$ (17% above threshold).

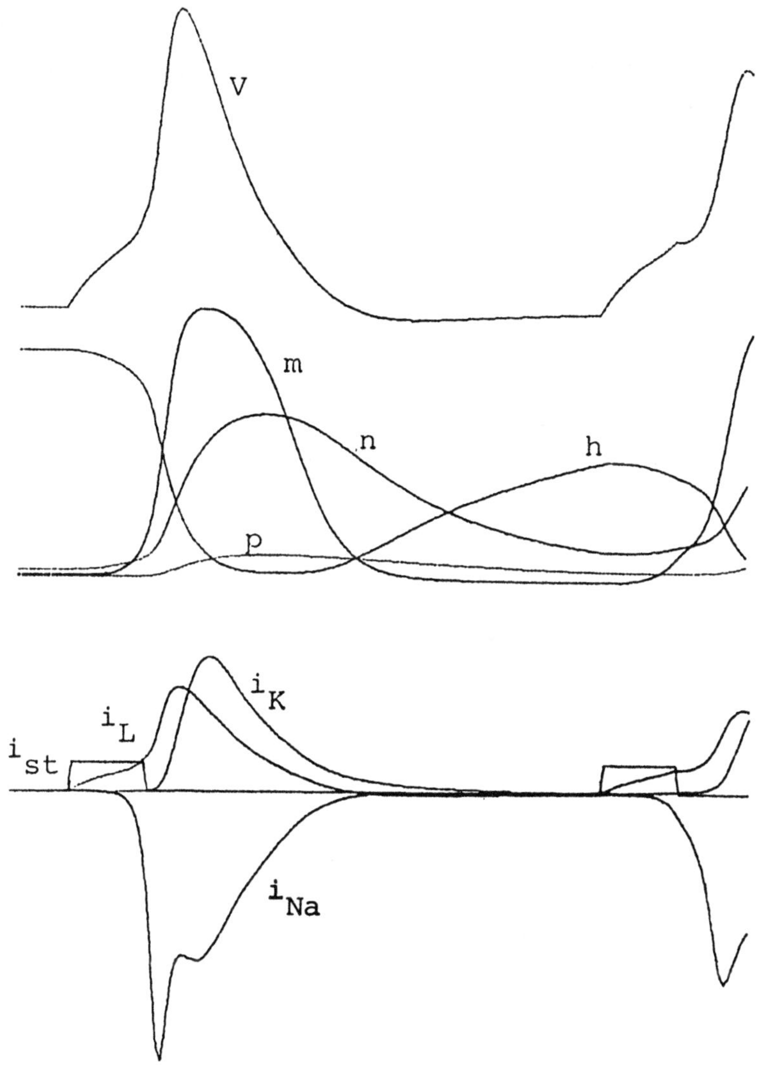

Fig. 5.14 *Refractory behavior of the FH model. The gating variables n, h and p react slower than in the CRRSS model. Thus, relatively stronger stimuli allows for an effective excitation when they are applied in at least 0.7 ms intervals. Although both models have similar AP durations, the FH model is characterized by longer refractory periods and higher thresholds.*

Simulation with standard data, T=37°C, Q_{10} from Table 5.2. Ipulses: 100μs, 900μA/cm^2 (24% above threshold).

good agreement with the experiment. The missing values can be found by data fitting of the measurements presented in their figures. By recalculation we can find similar values as obtained by SWEENEY *et al.* (1987), who transformed the original data by multiplying all the α's and β's with a common $Q_{10} = 3$ to find a model for warm blooded axons. The CRRSS – MODEL, named after the investigators CHIU, RITCHIE, ROGERT, STAGG - and SWEENNEY describes the sodium and leakage membrane currents in mammalian myelinated fibers. Box 5.3 gives both the equations and the constants of the CRRSS model. The model for the original experiments, made at 14^oC, is the same but all the α's and β's have to be divided by $3^{2.3} = 12.51$ according to $Q_{10} = 3$. Besides the HH and FH models SWEENEY's data are the most used for modeling medical applications of functional electrostimulation.

In contrast to the unmyelinated fibers, the subthreshold part of the models for myelinated fibers shows another type of behavior. This is caused by the high membrane capacity in the nodal area. Thus, the foot of an AP, generated by a current pulse, is not nearly linear (Fig. 4.6) but shows the behavior as predicted from the simple RC-circuit in Fig. (C) of Box 3.2. This exponentially-curved starting phase can be observed in all the illustrations from Fig. 5.8 to Fig. 5.12. When threshold is reached in the CRRSS model, the gating variable m reacts very quickly with a jump to the value 1, but the sodium current comes down because the inactivity variable h drops. Sodium current does not find a second peak like that found in the HH and FH models.

The short duration of the AP is caused by the high temperature. Original model data were found on rabbit nerves at a temperature of 14^oC. At this temperature, the shape of the AP as well as the duration (about $4ms$) is close to that of frog nerves. (CHIU *et al.*, 1979)

Fig. 5.12 illustrates the membrane behavior near threshold. Similar to the FH model, the subthreshold response drops quickly down to the resting voltage, however, hyperpolarization is not seen. It is not possible to reach the flat parts which can be obtained in the HH model as shown in Fig. 5.10 (B).

Due to the quick gating processes in the CRRSS model, there is a very short refractory period. Fig. 5.13 demonstrates that the membrane is able to fire with 2 kHz when stimulated by $100\mu s$ current pulses which are only 1.17 times stronger than threshold.

BOX 5.4	THE SCHWARZ-EIKHOF MODEL

The SCHWARZ-EIKHOF MODEL describes the voltage-current relations for membranes in the nodes of Ranvier in a rat moter nerve at 37^oC. The structure of the model is similar to that of FRANKENHAEUSER & HUXLEY (Compare Box 5.2, Table 5.2):

$$\dot{V} = [-I_{Na} - I_K - I_L + I_{st}]/C \qquad \text{(SE-1)}$$

with the ionic currents

$$I_{Na} = P_{Na} m^3 h \frac{EF^2}{RT} \frac{[Na]_o - [Na]_i e^{EF/RT}}{1 - e^{EF/RT}} \qquad \text{(SE-2)}$$

$$I_K = P_K n^2 \frac{EF^2}{RT} \frac{[K]_o - [K]_i e^{EF/RT}}{1 - e^{EF/RT}} \qquad \text{(SE-3)}$$

$$I_L = G_L(V - V_L) \qquad \text{(SE-4)}$$

where membrane voltage E is given by the reduced voltage V and the inside resting potential V_{rest}

$$E = V + V_{rest} \qquad \text{(SE-5)}$$

The gating variables are defined by

$$\dot{m} = -(\alpha_m + \beta_m) \cdot m + \alpha_m \qquad \text{(SE-6)}$$

$$\dot{n} = -(\alpha_n + \beta_n) \cdot n + \alpha_n \qquad \text{(SE-7)}$$

$$\dot{h} = -(\alpha_h + \beta_h) \cdot h + \alpha_h \qquad \text{(SE-8)}$$

with the coefficients

$$\alpha_m = \frac{1.87 \cdot (V - 25.41)}{1 - e^{\frac{25.41 - V}{6.06}}} \quad \beta_m = \frac{3.97 \cdot (21 - V)}{1 - e^{\frac{V - 21}{9.41}}} \qquad \text{(SE-9)}$$

$$\alpha_n = \frac{0.13 \cdot (V - 35)}{1 - e^{\frac{35 - V}{10}}} \quad \beta_n = \frac{0.32 \cdot (10 - V)}{1 - e^{\frac{V - 10}{10}}} \qquad \text{(SE-10)}$$

$$\alpha_h = -\frac{0.55 \cdot (V + 27.74)}{1 - e^{\frac{V+27.74}{9.06}}} \qquad \beta_h = \frac{22.6}{1 + e^{\frac{56-V}{12.5}}} \qquad \text{(SE-11)}$$

The initial conditions are

$$V(0) = 0, \quad m(0) = 0.0077, \quad n(0) = 0.0267, \quad h(0) = 0.76 \qquad \text{(SE-12)}$$

The constants are

$V_{rest} = -78mV$, $V_L = 0mV$, $[Na]_o = 154mmol/dm^3$, $[Na]_i = 8.71mmol/dm^3$, $[K]_o = 5.9mmol/dm^3$, $[K]_i = 155mmol/dm^3$, $R = 8314.4mJmol^{-1}K^{-1}$, $F = 96485Cmol^{-1}$.

The following constants depend on nodal area A, which we can assume to be in the magnitude of 50 μm^2=$0.5 \cdot 10^{-6} cm^2$ *

$P_{Na} = 0.00164 \cdot 10^{-9} dm^3 s^{-1}$, $P_K = 0.000067 \cdot 10^{-9} dm^3 s^{-1}$, $C = 1.4pF$ and $G_l = 43nS$

The model data use the temperature T=37^oC. Colder experiments can be simulated using $Q_{10} = 2.2$ for α_m and β_m, $Q_{10} = 2.9$ for α_h and β_h, $Q_{10} = 3.0$ for α_n and β_n.

---

* With this scaling we obtain the nodal currents in pA, voltage in mV, time in ms. In order to obtain current densities in $\mu A/cm^2$, as used in the other membrane models, the values for P_{Na}, P_K, C and G_l have to be doubled for A=50 μm^2.

By Fig. 5.14 it is shown, that the FH model reacts with a slightly longer refractory time and that higher stimuli are necessary for high firing rates.

In 1987 SCHWARZ & EIKHOF published a model developed from voltage clamp experiments on rat nodes (See Box 5.4). This model is of special interest for medical applications because most of the data was measured at 20^oC and at 37^oC. In 1986 SCHWARZ showed that sodium activation and inactivation react with a Q_{10} of about 1.8-2.1, for temperatures between 20^oC and 40^oC, with 2.6-2.7 for temperatures between 10^oC and 20^oC, and with Q_{10}'s up to 3.7 below 10^oC. However, leakage current is hardly affected

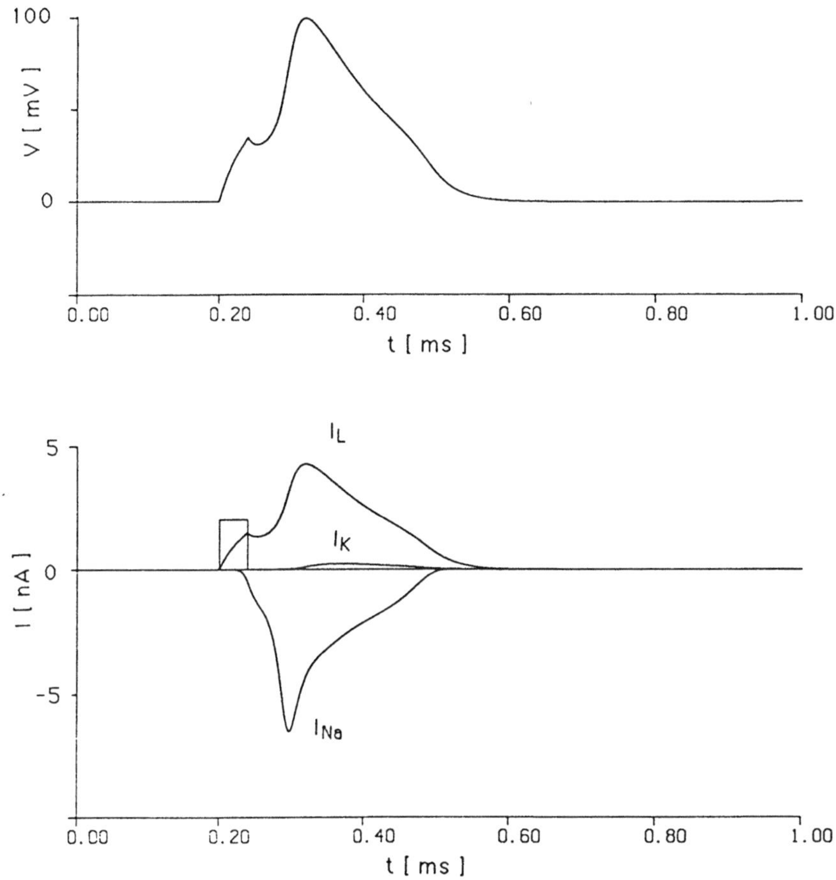

Fig. 5.15 Schwarz-Eikhof simulation of a rat motoneuron at 37°C. Stimulation with a 2nA injected current pulse of 40 μs duration.

by temperature and, as mentioned above, potassium current can be neglected in the intact fiber.

Fig. 5.15 shows the reaction of the SCHWARZ-EIKHOF MODEL when stimulated with a current impulse. The distribution of the ionic currents is similar to that of the CRRSS model (Fig. 5.11). The weak potassium current is neglectable. Fig. 5.15 gives the nodal currents. Assuming a nodal area of 50 μm^2, as in Box 5.4, we find that a current density of 4 mA/cm^2 results for the 60 μs stimulating pulse. Comparison of stimuli strengths of the models for unmyelinated fibers show that the FH axon has the smallest thresholds, the CRRSS model needs more and the SE

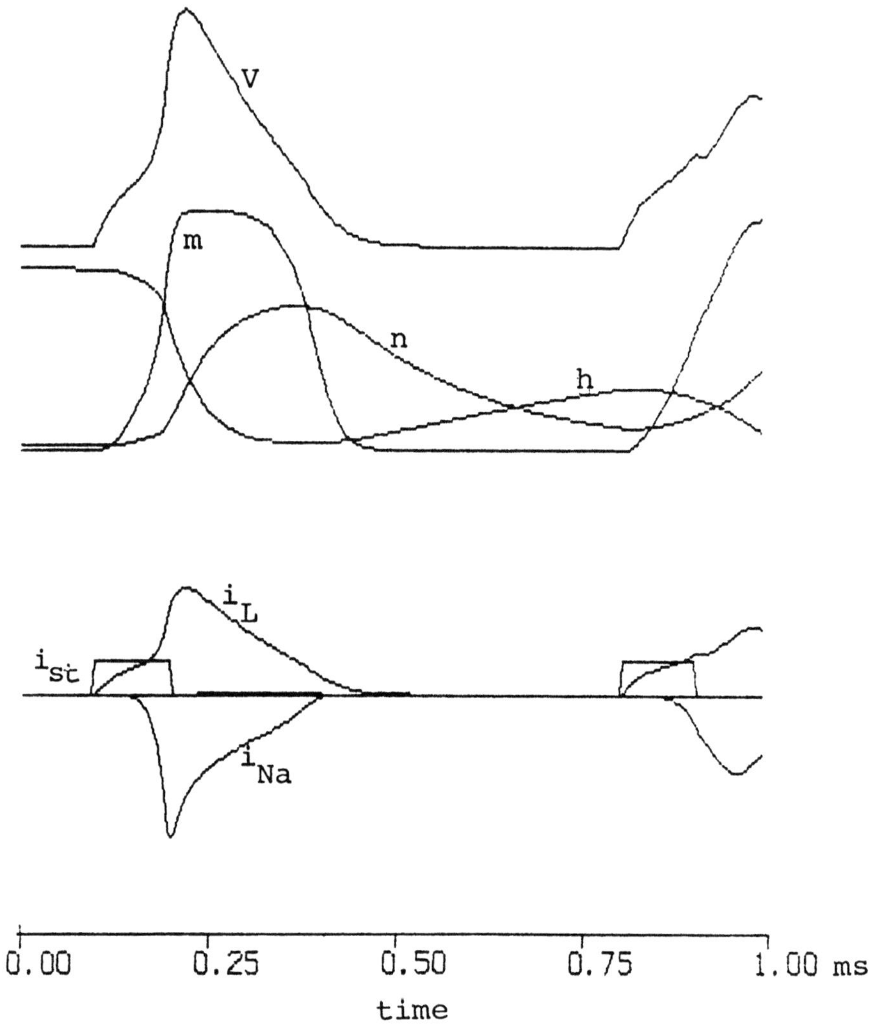

Fig. 5.16 Refractory behavior of the SE model. A second stimulus, 25% above threshold (2750 $\mu A/cm^2$, 100 μs) is applied 0.7 ms after the first.

model predicts the highest values.

Comparing the refractory behavior (Fig. 5.13, Fig. 5.14, and 5.16), we see that the modified FH model allows firing rates up to about 1.4 kHz, and the CRRSS model can fire with 2 kHz. Because the gating variable h reacts slowly in the SE model, the maximum firing rate is smaller than in the modified FH model.

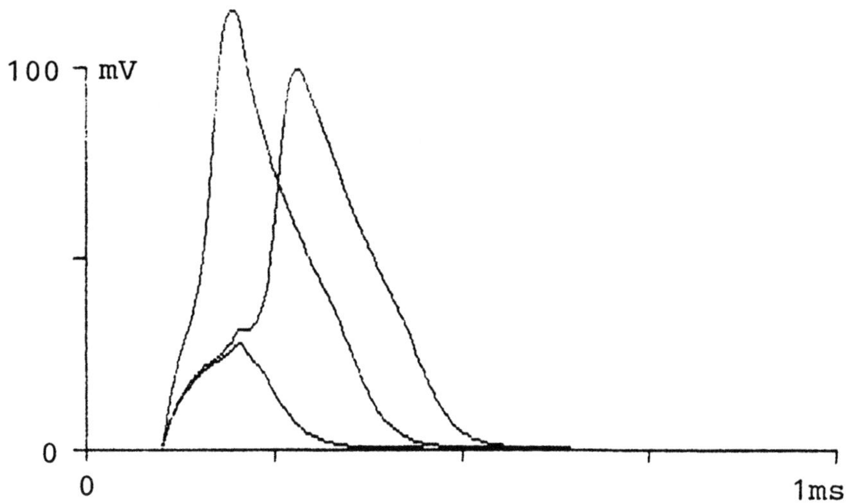

Fig. 5.17 *The reaction of the SE-model to pulses of different strengths is similar to the FH model, but without hyperpolarization. Current strengths: 2100, 2200, and 4000 $\mu A/cm^2$; pulse duration 100 μs.*

This is demonstrated in Fig. 5.16, where a simulated double pulse experiment (25 % above threshold, pulse intervals of 0.7 *ms*) will generate a considerably reduced second AP.

Fig 5.17 shows the reactions of the SE model to diffferent stimulation strengths. The behavior is similar to that of the modified FH model and to the CRRSS model.

There exist even more membrane models. They are concerned with special membranes and some are interesting because of the mathematical investigations. [For more details and references see, e.g., HOLDEN, 1980 or JACK *et al.* 1983.]

BEELER & REUTER (1974) gave equations for the fibers of ventricular muscle. MCALLISTER, NOBLE & TSIEN (1975) found a model for cardiac Purkinje fibers using six different ionic currents. CONNOR & STEVENS (1971) and CONNOR *et al.* (1977) developed models for repetitive firing. SCRIVEN (1981) also investigated in this field, modeling the ionic pumps, as well.

The differences in membrane reactions are caused by the number and the properties of ionic channels involved. Most excitation processes are carried out by only a few types of channels; therefore, we can simulate most cases with relatively small models of the HH type, where the less important ions are summarized into the nonspecific leakage current. From the view of electrostimulation, we are not interested in the knowledge of the ionic currents. But we are interested in the timing and threshold behavior of stimulated fibers.

6. PROPAGATION OF THE SPIKE

Electrical network to simulate fiber properties

The previous methods like the space-clamped experiments can be used to simulate the reactions of fibers with long inserted electrodes. But with those models we can also predict the voltage across the membranes of spherical cells, especially when we know the 'injected' stimulating current coming from the noninsulated tip of an inserted electrode. We find a different situation if we want to stimulate a nerve or muscle fiber electrically. Since the injected current leaks away on both sides of the fiber, the membrane voltage becomes a function of distance. Therefore, the situation is different from that of the space clamp, and each part of the fiber will react individually. We will overcome this difficulty by segmentation of the fiber into small cylinders with a length Δx. Within one segment the membrane behavior is similar, hence both voltage and currents may be appoximated by a mean value. Now, every segment of the fiber represents a space clamp experiment, and these can be modeled by an electric circuit, as shown in Fig. 6.1. A single segment will be simulated as a 'local model', e.g., with the HH equations or with any other model, as described in Chapter 5.

The electrical network of Fig. 6.1 can be used for myelinated and unmyelinated axons, as well as, for muscle fibers. In myelinated fibers, only the nodal part of length L is active. For myelinated fibers, the segmention length Δx is the internodal distance. However, in the case of unmyelinated fibers, Δx (and $L = \Delta x$) is determined by computational accuracy.

The current flow for the nth segment of the fiber, at the point marked with a full circle in Fig. 6.1, is caused by the voltage between the different points of the network. This consists of a capacitance current, different ionic currents, a current along the inside (defined by the inneraxonal conductance G_a and the voltage to the neighbor points), and eventually an injected stimulus current I_{st}

$$C_m \frac{d(V_{i,n} - V_{e,n})}{dt} + I_{i,n} + G_a(V_{i,n} - V_{i,n-1}) + G_a(V_{i,n} - V_{i,n+1}) = I_{st,n} \qquad (6.1)$$

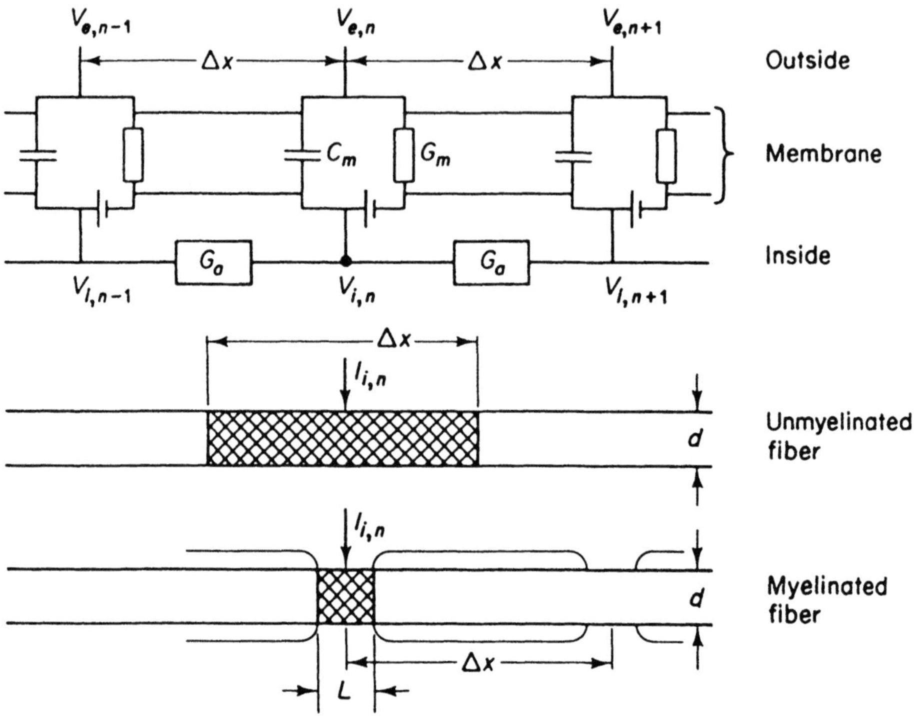

Fig. 6.1 *Electrical network to simulate the currents in a fiber. Unmyelinated, as well as, myelinated fibers of diameter d are segmented into cylinders of length Δx. Myelinated fibers have active membrane parts only in the cross-hatched area at the nodes of Ranvier. Here, ionic currents will only enter at a length L. The membrane of every cylinder is simulated by an electric circuit (top diagram) consisting of capacity C_m, voltage source, and nonlinear resistance. $V_{e,n}$ and $V_{i,n}$ are the external and the internal potential at the nth segment. G_m symbolizes the nonlinear membrane conductance and G_a the conductance of axoplasm between two segments. $I_{i,n}$ is the ionic current passing the membrane of the nth segment.*

(From Rattay, 1987 b)

In order to solve the system of equations 6.1 we have to add a subsystem of equations for the ionic currents (e.g. the HH equations) for every node. Furthermore, an assumption is needed on the extracellular potential V_e. If the fiber is immersed in a large volume of extracellular fluid the extracellular resistance may be neglected. This simplification was used by the pioneers when they examined the propagating AP, hence we will also set

$$V_e = 0 \tag{6.2}$$

for all the segments. CLARK & PLONSEY (1968) and PLONSEY (1974) examined the potential distribution on both sides of the membrane and found that only changes, in the order of some mV, occur on the extracellular side when an AP is passing. However, about 97% to 99% of the voltage is carried by the inside of an isolated fiber. Therefore, we can neglect the extracellular potential for simplified cases, as long as, no extracellular stimulation is applied. A mathematical description of the general case is given in WOOSLEY *et al.* (1985) and in ROTH & WIKSWO (1985), where the electromagnetic influences in a nerve bundle are also formulated.

With these basic assumptions (eqns. 6.1 and 6.2), we can solve plenty of different applications when propagation effects are involved. The simplest is that of injecting a current pulse locally in a fiber (Fig. 6.2). For this purpose, we use a system of equations derived from (6.1) by neglecting the extracellular potential and by introducing the reduced voltage, as it was used in the local models of Chapter 5:

$$V_{e,n} = 0 \quad \text{and} \quad V_n = V_{i,n} - V_{rest}$$

We assume that stimulus current is injected at $n = 0$. Since the reaction of the fiber is symmetric, $V_{-n}(t) = V_n(t)$. Thereby, the model is reduced to a system of N equations of the form

$$\dot{V}_0 = [-I_{i,0} + G_a(2V_1 - 2V_0) + I_{st}]/C_m \tag{6.3}$$

$$\dot{V}_n = [-I_{i,n} + G_a(V_{n-1} - 2V_n + V_{n+1})]/C_m \qquad \text{for } n = 1, 2, \ldots N-1 \tag{6.4}$$

and N subsystems (e.g. HH equations) for the ionic currents $I_{i,n}$. The variable V_N in the last equation (6.4) is not defined and it may be set to 0 or to V_{N-2} which simulates a closed or a reflecting end, respectively. Fig. 6.2 shows that the shape of the AP is the same in subsequent segments, but the AP's now appear with time delays. Therefore, the solutions for short times are not influenced by the choice of V_N.

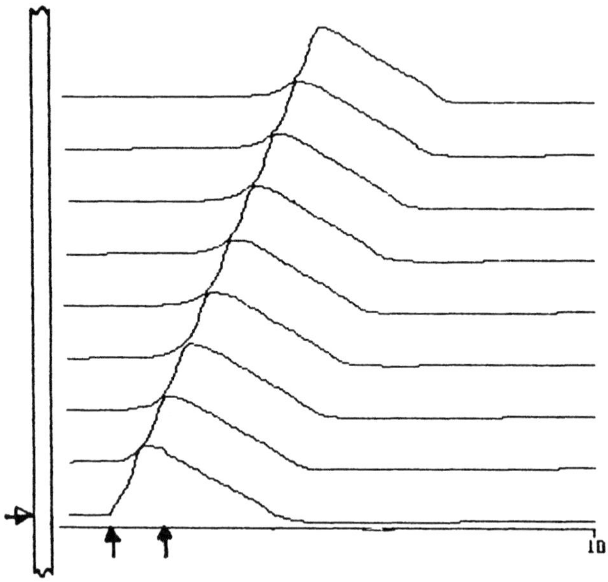

Fig. 6.2 *Propagation of an action potential. On the left side the fiber is marked with an arrow, where, from the top of an electrode, current is injected into the fiber. A square pulse of 1 ms (switching on and off time is marked by vertical arrows) produces an AP which propagates symmetrically to both sides, however, only the reactions of the upward part of the fiber are shown here. The lines correspond to points of the axon with a distance of 5 mm. The shape of the propagated spike is very close to that obtained by solving a single HH model (for the space clamped case).*

Simulation was done with eqns. 6.3 and 6.4 according to the network of Fig. 6.1 with HH standard data as in Table 4.1 and fiber diameter d=0.0048cm, axial conductance $G_a = 0.0515 mS.cm$, *and a simulation time of 10ms.*

(From Rattay, 1987a)

A very similar result to that of Fig. 6.2 is obtained in the case of natural stimulation when a superthreshold voltage is produced at the receptor cell. Because receptor cells or signal-sending parts are found only at one end of an axon in natural stimulation, signals are sent in one direction only. This is in contrast to the symmetric excitation obtained with electrostimulation.

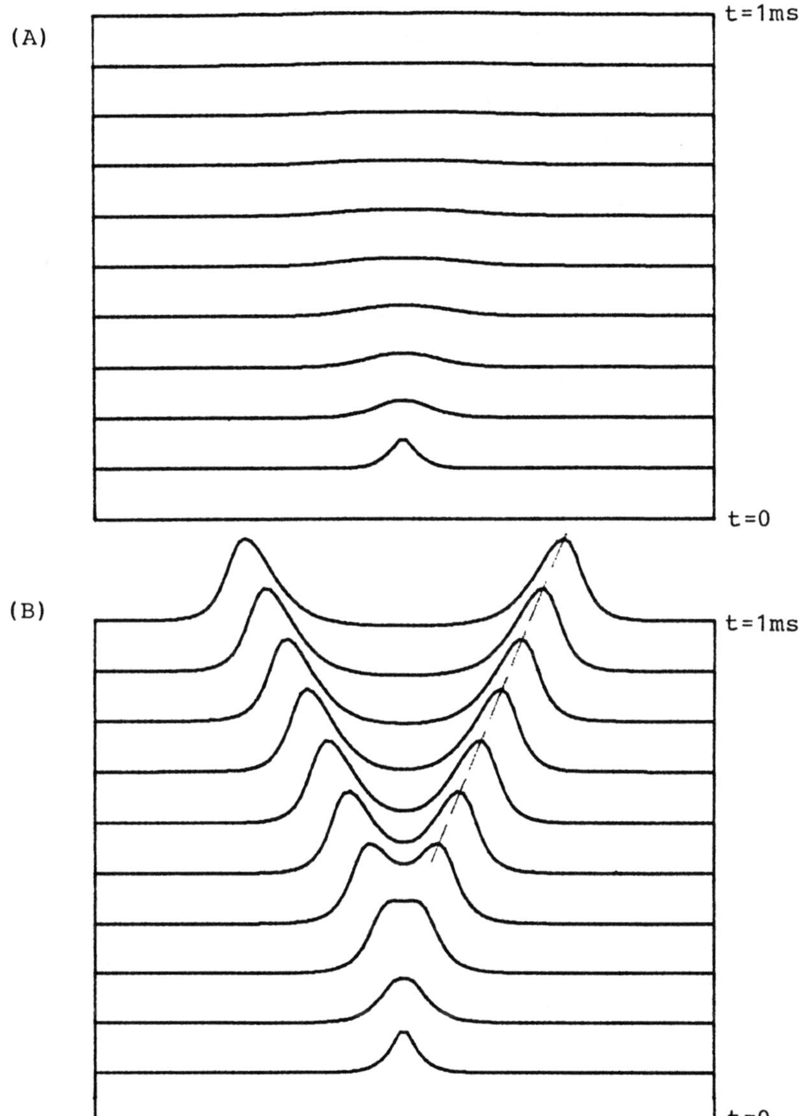

Fig. 6.3 Snapshots of membrane voltage along the unmyelinated axon for t=0, 0.1, ... 1 ms. (A) subthreshold (B) superthreshold responses to a 100μs *voltage distribution injected current square pulse at x=0. Every line gives the symmetric voltage distribution along the axon for a fixed time. After bifurcation, the AP propagates to both sides with a constant velocity (dashed line). Simulation is with standard HH data, but for* $T = 29^oC$, *the axon diameter is* 10μ, *and stimulation current amplitude is* 3.3nA *for (A) and* 4.4nA *for (B), 100 segments are used to obtain a 'continuous' spatial reaction.*

Voltage and current distribution along the axon

If we stop the calculation of the system (6.3) and (6.4) at a special time we get a snapshot of the voltage along the axon. This is illustrated in Fig. 6.3 for subthreshold and superthreshold stimuli. A $100\mu s$ current pulse is injected at x=0 into a $10\mu m$ unmyelinated axon of the HH type. A part of the injected current leaks away along the axis of the axon and lifts up the voltage there [lowest curve of (A) and (B)]. After $300\mu s$, the amplitude of the AP is reached and the peak bifurcates [Fig. 6.3 (B)]. Now, the AP gets its characteristic form and propagates with the constant velocity indicated by the dashed line in Fig. 6.3 (B). The highest trace shows the result after 1 ms. The left peak looks similar to the time response, known from Fig. 6.2, and from the space clamp experiments (Chapter 4). Differences between the shapes of the AP's are caused by different temperatures. Indeed, experiment and calculation show that the hump in the falling part of the AP disappears with rising temperatures.

Because the equations for the ionic currents are formulated with current densities, it is convenient to transform (6.1) by setting

$$G_a = \frac{\pi d^2}{4\rho_i \Delta x} \qquad \text{and} \qquad C_m = c_m \pi d L \tag{6.5}$$

where: ρ_i is the specific resistance of axoplasm which is about $0.1k\Omega.cm$, c_m is the capacity per cm^2 of free membrane, and L is the nodal gap width. We can treat the case of the myelinated and the unmyelinated fiber in parallel if we set $L = \Delta x$ for unmyelinated fibers. Equation 6.1, together with practical physical units becomes

$$\begin{aligned}\frac{d(V_{i,n} - V_{e,n})}{dt}\left[\frac{mV}{ms}\right] = \Bigg\{ & -i_{ionic}[\mu A/cm^2] \\ & + \frac{d[cm]}{4\rho_i[k\Omega cm]\cdot\Delta x[cm]\cdot L[cm]}\cdot(V_{i,n-1} - 2V_{i,n} + V_{i,n+1})[mV] \\ & + \frac{I_{st}[\mu A]}{\pi d[cm]L[cm]}\Bigg\} / c_m[\mu F/cm^2]\end{aligned} \tag{6.6}$$

The influence of fiber diameter on propagation velocity can be derived from (6.6) if $V_{e,n} = 0$. For this purpose, we assume that

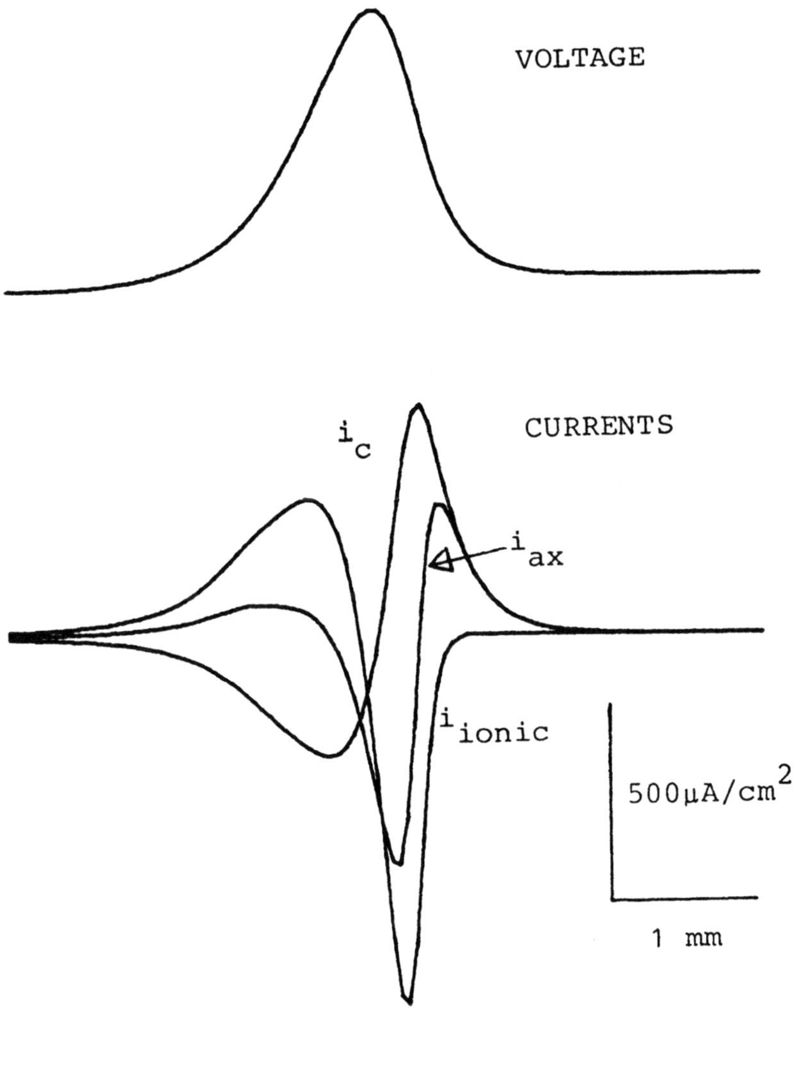

Fig. 6.4 *Voltage and current distribution along the axon. This figure is the simulation of a propagating AP along an unmyelinated axon as in Fig. 6.3, 1 ms after excitation, but it is only for the positive x. The voltage corresponds with the right part of the top trace of Fig. 6.3 (B). The lower curves show the distribution of the capacitive current i_c, the axial (axoplasm) current i_{ax} and the ionic current, per cm^2 of membrane involved. The propagation of an AP is carried by the exponential decay of i_{ax} which is the subthreshold response of the right part of the axon. This current loads the capacity of the membrane, thereby producing an exponential leading edge of the AP.*

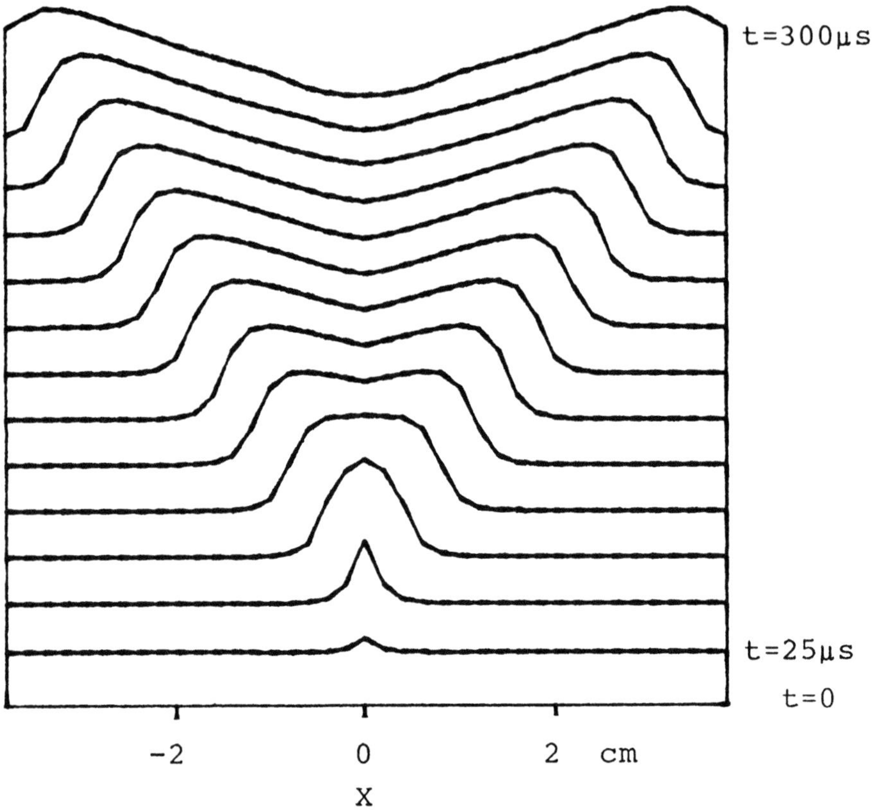

Fig. 6.5 Voltage distribution along a myelinated fiber at different times. 8 cm of the axon are segmented with $\Delta x = 2mm$ according to the nodal distance. The values for the voltage are defined only at those discrete points which are seen by plotting weakness (points are connected by linear interpolation). The result is similar to that of Fig. 6.3 but high velocity is caused by myelin sheet.

Simulation was done with the model of Box 5.3 for a $10\mu m$ fiber diameter, and a nodal gap width of $L = 2.5\mu m$. Stimulation with a monopolar, extracellular electrode located 1 mm above the node $x = 0$. Axoplasm resistivity $\rho_i = 0.055k\Omega.cm$, extracellular resistivity $\rho_e = 0.3k\Omega.cm$, stimulus pulse strength $-30\mu A$, duration 0.1 ms.

the solution of (6.6) is computed for an unmyelinated fiber ($L = \Delta x$) of diameter d at the N supporting points $x_n = n.\Delta x$. We consider now a propagating AP which is far away from the point of generation, and therefore, it is not influenced by the stimulus

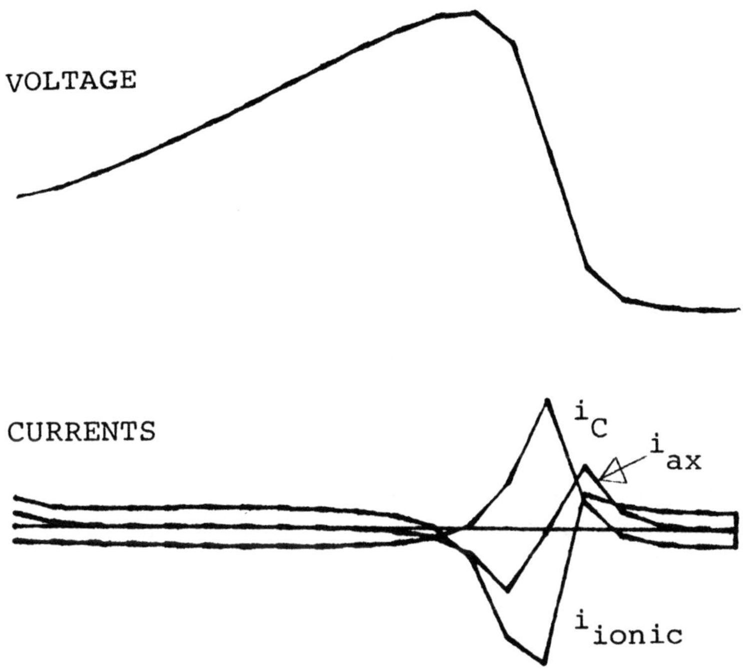

Fig. 6.6 Voltage and current flow along a myelinated fiber. The situation of Fig. 6.5, at t=250 μs, for the 20 nodes at x=0, 2, ... 38 mm. The current through the membrane ($i_c + i_{ionic}$) is equal to the axial current i_{ax} at every node.

signal. If we consider another fiber with diameter $d_1 = k \cdot d$ and if we make a new choice for $\Delta x_1 = \sqrt{k} \cdot \Delta x$, we would get exactly the same coefficients in (6.6). Therefore, we get the same results at the points of support at the same times as in the first case, but now at the distances $\sqrt{k} \cdot \Delta x$. This means that velocity of propagation is proportional to the square root of diameter in unmyelinated fibers.

In the case of myelinated fibers, we will assume that the nodal distance Δx is proportionate to d, but that the gap width L is independent of d. [Compare, e.g., RUSHTON (1951) or JACK *et al.* (1983) for more detailed examinations.] In this case, a changed diameter $d_1 = k \cdot d$ would demand for $\Delta x_1 = k \cdot \Delta x$ and again we obtain the same values at the supporting points. This means that velocity is proportional to diameter. This is in agreement with the empirical rule

$$\text{velocity in [ms] is 4.5 times } d \text{ in } [\mu m] \tag{6.7}$$

for human applications.

Furthermore, it is seen with these assumptions that the duration and even the shape of the AP is independent of the diameter for both the myelinated and the unmyelinated fiber.

In a similar way, we see the influences of ρ_i or L on conduction velocity. For instance, the results of Fig. 6.5 were found with a relative low $\rho_i = 0.055k\Omega.cm$ and an evaluation of the figure shows a velocity of about 135 m/s instead of 45 m/s. Multiplication of ρ_i or L by a factor 3 gives the empirical result. Of course, a combination of changed parameters is also possible, e.g., $\rho_i = 0.083k\Omega$ and $\Delta x = 1mm$ (which correspond to a more realistic case) satisfies (6.7) for $d = 10\mu m$.

The propagation in a $10\mu m$ unmyelinated fiber is much slower. In the case shown in Fig. 6.3 (B), the velocity is about $3.5m/s$. Even for a $490\mu m$ diameter of giant squid axon the 7 fold velocity (24.5 m/s) is slow compared with that of myelinated axons. For the assumptions made above, we get the following velocities*:

$$v_{myelinated} = 4.5 \cdot d \tag{6.7}$$

and

$$v_{unmyelinated} = 1.1 \cdot \sqrt{d} \tag{6.8}$$

Comparing these equations, we see that in very small fibers ($d < 0.25\mu m$) AP's propagate quicker within unmyelinated axons than within myelinated ones (Fig. 6.7). [With other assumptions the diameter at which unmyelinated axons are quicker is even larger; RUSHTON, 1951.]

---

* Another rule of the thumb says that for fibers thicker than 11μ, $v_{myelinated} = 6 \cdot d$

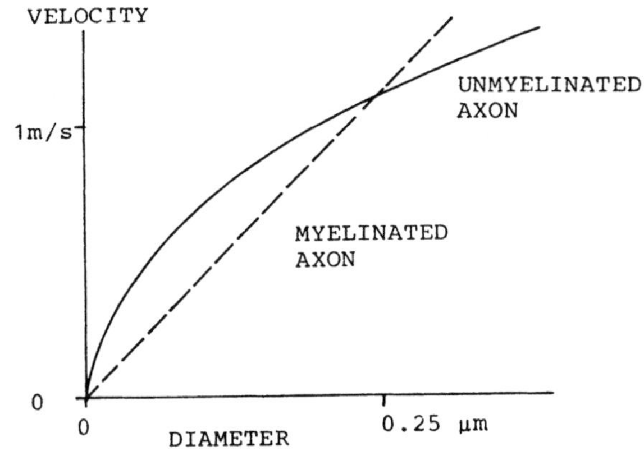

Fig. 6.7 Propagation velocity of small axons as a function of diameter. Conduction velocity of nerve signals is proportional to the diameter in myelinated fibers (dashed line) and is proportional to the square root of diameter in unmyelinated axons. At a special diameter, (with the assumptions made in text, this diameter is about 0.25µm) both types of axons have the same propagation velocity. For smaller diameters myelinated fibers are slower.

The cable equation

The discrete version of (6.6) is especially suitable for myelinated fibers. For analytical considerations of unmyelinated axons, it is often more instructive taking the limit $\Delta x \to 0$ and without an injected stimulus (6.6) reads as

$$\frac{\partial(V_i - V_e)}{\partial t} = [-i_{ionic} + \frac{d}{4\rho_i} \cdot \frac{\partial^2 V_i}{\partial x^2}]/c_m \tag{6.9}$$

where V_i, V_e and i_{ionic} are functions of x and t. By setting $V_e = 0$, $V_i - V_{rest} = V$, $\frac{d}{4\rho_i} = g_a$ and $i_{ionic} = g_m.V$ eqn. 6.9 may be written in the simplified form

$$g_a \frac{\partial^2 V}{\partial x^2} = g_m V + c_m \frac{\partial V}{\partial t} \tag{6.10}$$

For constant g_m (6.10) is called the 'cable equation' because it describes the voltage in long electric cables. We can use the cable equation for subthreshold analysis.

If the axon is stimulated by a constant subthreshold current, injected at $x = 0$, the membrane conductance g_m may be assumed to be constant. With this constant current we observe that the steady state solution $[\frac{\partial V}{\partial t} = 0]$ for all points along the axon and (6.10) becomes

$$g_a \frac{\partial^2 V}{\partial x^2} = g_m V.$$

We set $\frac{g_a}{g_m} = \lambda^2$ and obtain an exponential decay at both sides of the axon:

$$V(x) = V(0) \cdot e^{-\frac{|x|}{\lambda}}$$

λ is called the **space constant**. It gives the distance where V falls to $\frac{V}{e}$, i.e., V loses 63% of its value. λ is in the order of some mm and even smaller in smaller nerves. This means that subthreshold signals vanish within short distances [see also Fig. 6.3 (A)].

The space clamp reaction after a subthreshold stimulus can also be approximated by (6.10). Here we get exponential decay again. Setting

$$\frac{\partial^2 V}{\partial x^2} = 0$$

and assuming constant g_m (6.10) reads as

$$0 = g_m V + c_m \dot{V}$$

Plus setting $c_m / g_m = \tau$, we obtain

$$V(t) = V(0) \cdot e^{-t/\tau}$$

The **time constant** τ gives the time when subthreshold excitation loses 63% of its value (conf. also Box 3.2). Thus τ and λ characterize the subthreshold decay in the time and space domain.

The propagating edge

In contrast to the space clamp experiment, a firing axon generates excitation along the axon by itself and the stimulus signal is $g_a \frac{\partial^2 V}{\partial x^2}$ instead of the injected stimulus current i_{st}. Although stimuli signals could be arbitrary in functional electrostimulation, square pulses are preferred because they produce a linear foot at the exciting point similar to the space clamp excitation (first line of Fig. 6.2). The influence of the shape of the stimulus signal on the leading edge of the AP potential vanishes quickly and an exponential rise is seen in the shape of the spikes at a further distance away (upper curves in Fig. 6.2).

Instead of the injected current, the inneraxonal current i_{ax} stimulates the resting part of fibers when the spikes are propagating (Fig. 6.4). i_{ax} builds up an exponential rise at the propagating leading edge. This loads the capacity of the membrane. While extracellular potential is held at 0 V, (currents through the membrane leak away into the extracellular fluid) the capacitive current produces a voltage response. When threshold is reached, gating mechanisms become active and ionic current flow starts with an exponential rise of sodium influx (Fig. 6.4).

Case studies with currents applied at the inside: heat block, collision block, inverse stimulation

Stimulation with intracellular currents are not practicable in medical applications. Nevertheless, it is instructive to study propagation and blocking phenomena elicited by currents applied at the fiber's inner linings.

In the first case, we assume that the stimulating current enters an axon, 10 μm in diameter, from the tip of an electrode inserted at $x = 0$. Fig. 6.8 shows the reaction of a HH fiber under different temperatures when stimulated from such an inside point source. The fiber is parted in 0.01 cm segments, and only the central segment has a 100μs impulse, with a current density of 1500$\mu A/cm^2$, applied. With the help of eqn. 6.6, we can calculate the equivalent electrode current I_{st}

$$1500\mu A/cm^2 = \frac{I_{st}}{\pi dL} = \frac{I_{st}}{\pi \cdot 0.001 \cdot 0.01}$$

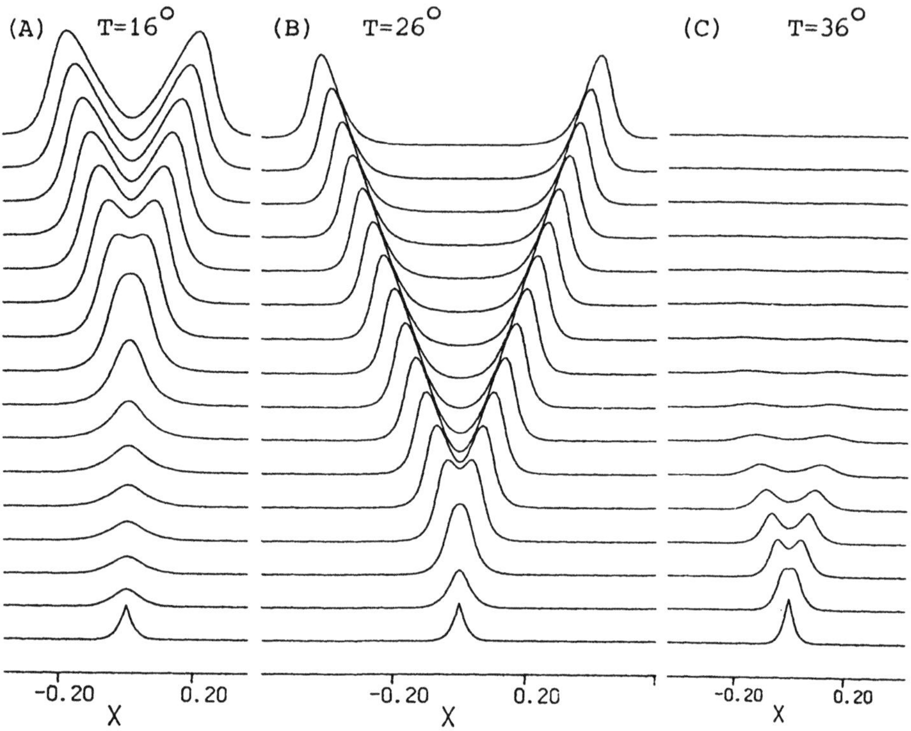

Fig. 6.8 Influence of temperature. The increase of temperature makes stimulation easier (B), but at high temperatures (C) the excitation will not propagate (heat block).

Simulation with standard HH data, but $d=10\mu m$, and $\Delta x=0.01$ cm. A $100\mu s$ current impulse of $1500\mu A/cm^2$ is applied only in the central segment at $x=0$. From the 200 segments used for calculation only the central part (approx. 0.8 cm) is shown.

and we find $I_{st} = 4.71nA$. Note, that this local stimulus current strength ($1500\mu A/cm^2$) is much higher than that known from the space clamp experiment (Fig. 4.6), because in our case only a relatively small part of the current (radial current) drives the membrane locally. Most of the current runs away down both sides inneraxonally (longitudinal current).

The lowest line of Fig. 6.8 shows the reaction of the membrane along the axon at the end of the stimulating pulse ($t = 100\mu s$). Higher peaks at higher temperatures indicate that excitation becomes easier with increasing temperature. Also the bifurcation of the AP [propagating in both directions in (A) and (B)] arrives sooner at T=26^oC than at T=16^oC. Although this trend is seen

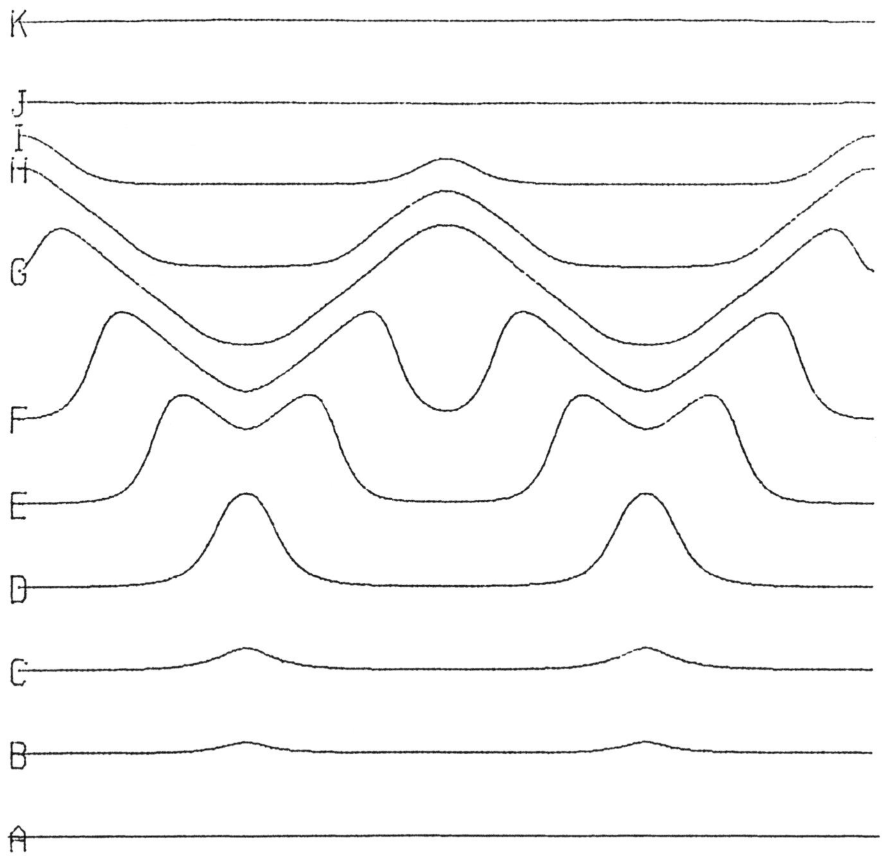

Fig. 6.9 Collision block. Two electrodes are used to generate two pairs of AP's. The AP's which propagate to the center produce a strong excitation when they come together, but since no bifurcation occurs it is concluded that when two AP's collide no one can travel through.

Simulation with standard HH data, $d=10\mu m$, $\Delta x=0.01cm$, and current strength equivalent to $50\mu A/cm^2$ is applied at 6 segments from $x=0.34$ to $x=0.39cm$ on both sides. Line A: resting state (start of impulse), line B: membrane voltage 0.5ms later, line C: reaction at the end of the 1ms stimulating pulse; simulated fiber length: 0.8cm.

in (C), no propagation will occur for high temperatures. This effect, which was already discussed in Chapter 4, is called **heat block**.

In the second example, we will simulate the blocking effect

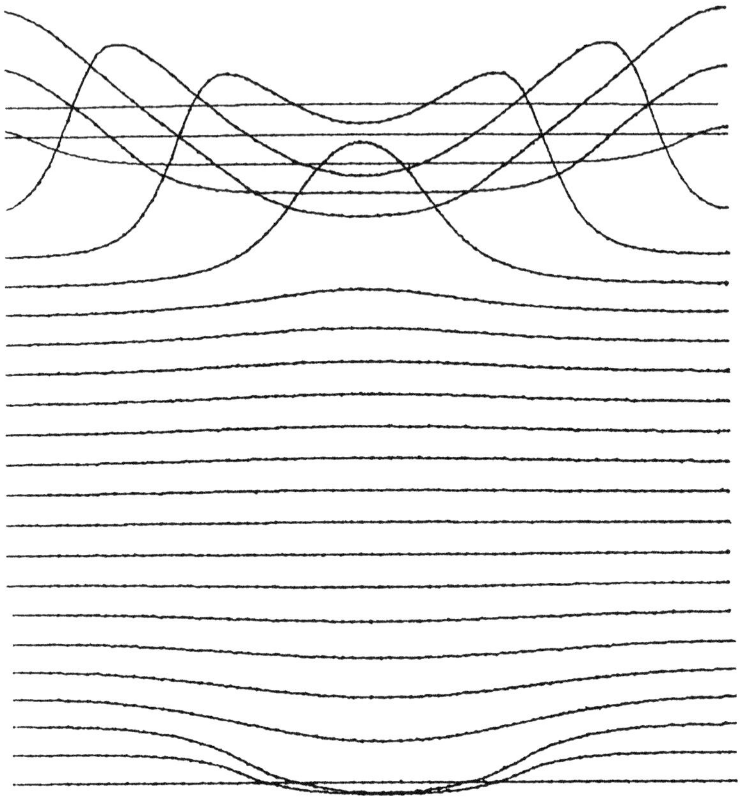

Fig. 6.10 Inverse inside stimulation. Strong currents applied at the central part of the axon will generate a late AP, as predicted from the local HH model by the swinging through phenomenon.

Computation is with standard HH data, T=6.3°C; but d=10μm, Δx= 0.01cm, and the pulse duration is 100μs. Current strength equivalent to 500μA/cm² is applied from x=-0.14cm to x=0.14cm. Reactions are displayed for 80 segments (0.8cm fiber length). Total simulation time is 12 ms, and the line spacing is 0.5ms.

caused by two AP's coming from different sides (Fig. 6.9). Two electrodes, in the form of non-insulated wires lying in the center of the axon, are used for stimulation. The electrodes are active in the segments 35 to 40 and -35 to -40. With $\Delta x = 0.01cm$, this means that stimulation is applied to a length of $0.6mm$ at both sides symmetric to $x = 0$. $1ms$ impulses, producing an equivalent of $50\mu A/cm^2$, are used to generate two AP's at symmetric positions. One could assume that less current is necessary compared to the first example, but now, the pulse duration is 10 times greater and furthermore, six segments are used on both sides. Therefore, the double charge is applied in the second example.

After $2ms$, the right peak bifurcates to two AP's which travel to different sides. At the left side there is the same situation and we get four AP's in line F of Fig. 6.9. Now, we are interested in the reaction produced by the collision of the AP's which move against each other. They produce a strong reaction (line G) at the meeting point, but no further bifurcation occurs and the peak comes down to the resting value (lines I and J). Due to the fact that it was not possible to overcome the hyperpolarization regions produced at both sides (line G), bifurcation has failed. The same situation takes place if the two AP's are coming from distant points. When they meet, neither AP can pass through. This phenomenon is called **collision block**. It is of special interest for medical application, for instance, to stop spastic signals by unidirectional firing: the artificially produced AP only travels in one direction and stops a naturally evoked spastic signal.

The last example shows that the propagation of an AP is also possible in the case of **inverse inside stimulation**. In contrast to the examples above, it is necessary to use a longer electrode. Here, the electrode is active at 29 segments where it is supplied with a strong negative impulse (Fig. 6.10). Due to the relatively long active electrode, the situation is similar to that of the space-clamp experiment. The sensitive reaction is not disturbed by currents flying away along the fiber.

7. EXTRACELLULAR STIMULATION OF FIBERS

The influence of the monopolar electrode on the extracellular potential

Several textbooks are concerned with the contents of Chapters 2-6, and more information can be found, in the books of:

AIDLEY 1979, CARPENTER 1984, COLE 1968,
HODGKIN 1964, JACK *et al.* 1983, KATZ 1966,
KEYNES & AIDLEY 1983, KUFFLER *et al.* 1984,
NOBLE 1979, RYALL 1979, SCOTT 1977 and WALSH 1964.

However, there is a lack of information about the theory on functional electrostimulation.

We will start our analysis of the external stimulation of axons with a simple practical example. A small spherical stimulating electrode is placed at a distance z close to an axon (Fig. 7.1). The ground electrode is relatively far away from this electrode. Thus, the reaction of the fiber will depend only on the distance z and on the shape of the stimulating signal. We assume that a current square pulse is used in order to stimulate the axon. The extracellular medium normally has a specific resistance which is in the order of $0.3k\Omega.cm$. Cell membranes around the electrode do not considerably hinder the sharp rising and falling edges of the signal used, therefore, the extracellular potential V_e may be approximated only by ohmic resistance:

$$V_e = \frac{\rho_e I_{el}}{4\pi r} \tag{7.1}$$

where $I_{el}(t)$ is the electrode current, and r gives the distance to the electrode. The extracellular potential is proportionate to $\frac{1}{r}$ and because $r = \sqrt{x^2 + z^2}$ (Fig. 7.1) we can find by eqn. (7.1) that the extracellular potential along the axon is

$$V_e = \frac{\rho_e \cdot I_{el}(t)}{4\pi\sqrt{x^2 + z^2}} \tag{7.2}$$

This relation can be confirmed in an experiment where a measuring electrode is moved, in a bath, along the x-axis. The same

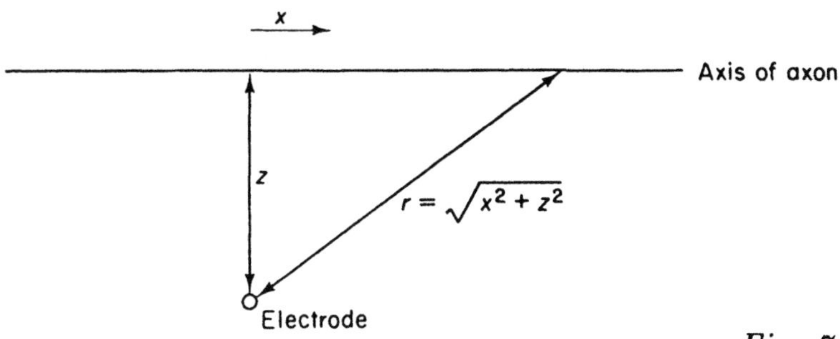

Fig. 7.1

experiment is especially helpful in order to find the influences of complicated shapes of electrodes.

We have mentioned in Chapter 6, that the influence of nerve's own activity on the extracellular potential is small and as long as not many fibers are active at the same time, the calculation of V_e can be done independently from both the geometry and the activity of fibers.

Now, the activity of the axon can be calculated at the points $x_n = n \cdot \Delta x$

$$\frac{d(V_{i,n} - V_{e,n})}{dt} = [-i_{ionic} + \frac{d}{4\rho_i \Delta x.L} \cdot (V_{i,n-1} - 2V_{i,n} + V_{i,n+1})]/c_m \tag{7.3}$$

which is derived from (6.6).

By introduction of reduced voltages as in (4.3)

$$V_n = V_{i,n} - V_{e,n} - V_{rest} \tag{7.4}$$

(7.3) reads as

$$\frac{dV_n}{dt} = [-i_{ionic} + \frac{d}{4\rho_i \Delta x.L} \cdot (V_{n-1} - 2V_n + V_{n+1} \quad + \\ V_{e,n-1} - 2V_{e,n} + V_{e,n+1})]/c_m$$

which can be written in the form

$$\dot{V}_n = \left[- i_{ionic} + \frac{d \cdot \Delta x}{4\rho_i \cdot L} \cdot \left(\frac{V_{n-1} - 2V_n + V_{n+1}}{\Delta x^2} + \right.\right. \\ \left.\left. \frac{V_{e,n-1} - 2V_{e,n} + V_{e,n+1}}{\Delta x^2} \right)\right] / c_m \tag{7.5}$$

Thereby, it is demonstrated that the stimulating influence of the extracellular potential is given in a segmented fiber by

$$\frac{V_{e,n-1} - 2V_{e,n} + V_{e,n+1}}{\Delta x^2} \tag{7.6}$$

which is the second difference quotient of the extracellular potential along the axon.

Equation 7.5 is valid for fibers both with and without a myeline coat. For unmyelinated fibers, $L = \Delta x$ and with $\Delta x \to 0$, we get

$$\frac{\partial V}{\partial t} = \left[- i_{ionic} + \left(\frac{\partial^2 V}{\partial x^2} + \frac{\partial^2 V_e}{\partial x^2} \right) \right] / c_m \tag{7.7}$$

The activating function

The classical cable equation (6.10) which was already used by the pioneers (COLE, 1968; HODGKIN & HUXLEY, 1952) is completed in (7.7) by the term $\frac{\partial^2 V_e}{\partial x^2}$.

$$f = \frac{\partial^2 V_e}{\partial x^2} \tag{7.8}$$

is a function of x and t. We will call f the **activating function,** because f is responsible for activating a fiber by extracellular electrodes. We will repeat the main principle of electrostimulation:

In unmyelinated fibers, a first approach of the influence of extracellular electrodes is given by the activating function which is the second derivative of the extracellular potential along the fiber.

With the help of the activating function, we can analyze different geometrical situations, but we can also compute a special situation using the following procedure

i) calculation of V_e as a function of x and t.
ii) calculation of the activating function (7.8).
iii) solving of (7.7), together with a system of differential equations for the ionic currents as described in Chapter 5.

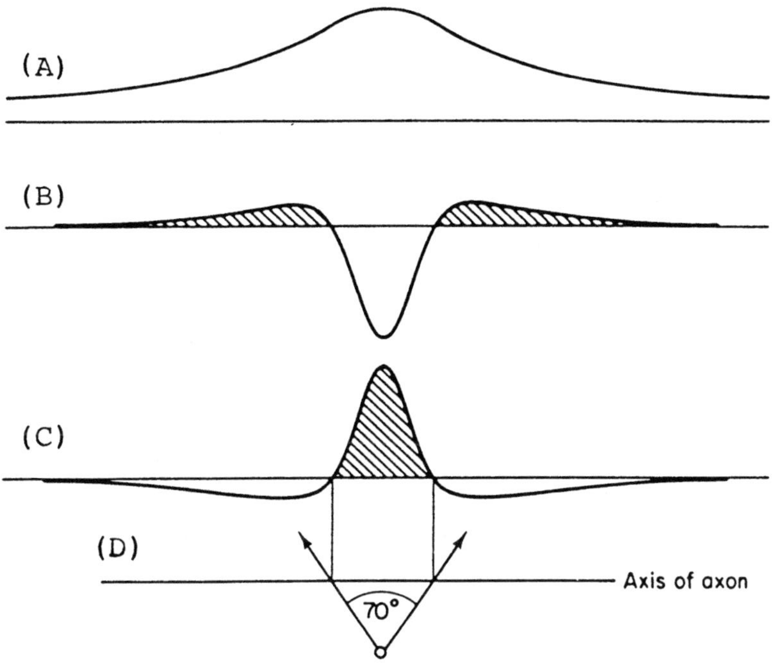

Fig. 7.2 Influence of a small stimulating electrode along a fiber. (A) shows the change of the extracellular potential for an anodic stimulation. Also the activating function is shown for anodic (B) and cathodic (C) stimulation. The horizontal axis corresponds to the x-axis, being the axis of the fiber. The parts where the fiber is depolarized are shaded. (D) shows the position of the electrode to get the upper traces. The border between the depolarizing and hyperpolarizing regions is given by an angle of about 70°. This angle does not depend on fiber parameters or on the conductance of the extracellular medium as long as it is homogeneous and isotropic.

(After Rattay, 1987b)

If the fiber was segmented with Δx, it is not necessary to calculate the activating function explicitly and instead of the steps ii) and iii) eqn. 7.5 can be used, together with equations for the ionic currents.

Now we will come back to our example of a single stimulating electrode. With the help of (7.2) we can obtain the activating function

$$f = \frac{\partial^2 V_e}{\partial x^2} = \frac{\rho_e I_{el}}{4\pi}(x^2 + z^2)^{-\frac{5}{2}}(2x^2 - z^2) \tag{7.9}$$

We assume that, at $t = 0$, no stimulus is applied and that the fiber is in the resting state, i.e., $V(x) = 0$. To reach an AP, depolarization is necessary (V has to become positive). As we start at $V = 0$ $\frac{\partial V}{\partial t}$ of (7.7) must be positive in the area where the excitation is to be produced. Therefore, the extracellular electrode stimulates the fiber in an area where $f > 0$. Figure 7.2 shows the extracellular potential V_e for a positive electrode current (A), and the activating function for anodic (B) and cathodic (C) stimulation. It is also known from experiments that the most excitable point of the axon, $x = 0$, may be stimulated with a cathodic current. This is because the positive part of f is strongest there (Fig. 7.2 (C)).

From (7.9) we find $f = 0$ at $x = \pm z/\sqrt{2}$. This means that a cathodic (negative) stimulus depolarizes the fiber over a length of $x = z\sqrt{2}$. This geometric relation requires an angle of 70.5°, to separate the activated and the deactivated parts of a fiber. This occurs when it is stimulated by a single monopolar electrode [cf. Fig. 7.2(D)].

Note that the polarizing area of the fiber is proportional to z; therefore, the threshold voltage across the membrane increases for short z. In the outside regions cathodic stimulation produces hyperpolarization. Strong cathodic stimulation hyperpolarizes the outside areas in such a way that a propagation of AP's is stopped there. This phenomenon will be discussed below.

Fig. 7.3 illustrates the effects of the activating function. An electrode, in the distance of 1 mm, stimulates an unmyelinated fiber with a cathodic 100 μs square pulse. After the end of the pulse, the voltage along the fiber is a picture of the activating function, known from Fig. 6.2 (C). We get this result because the subthreshold membrane conductance is nearly constant. That behavior allows us to simulate the subthreshold regions without using differential equations for the ionic currents. Thereby, computation time is considerably lower.

In his outstanding paper which has stimulated many investigators McNeal (1976) used this technique for analyzing excitation of myelinated nerves. He considered 11 points of support,

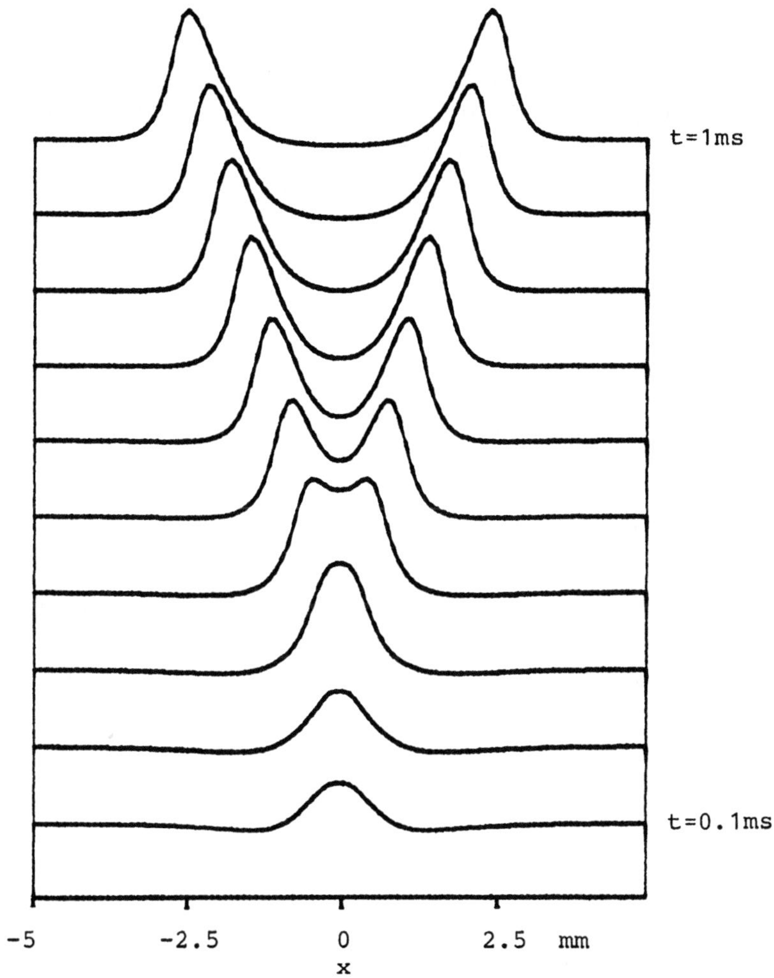

Fig. 7.3 *Voltage along an unmyelinated axon as a response to an extracellular cathodic stimulus signal. At the end of a* $100\mu s$ *square pulse, the distribution of the voltage is similar to the activiting function as a consequence of the nearly constant subthreshold membrane conductance. After threshold is reached the axon reacts in the same way as known from the current injection experiment shown in Fig. 6.3(B). Data from Fig. 6.3 but extracellular stimulation,* $I_{el} = -1.5mA$, $z = 1mm$, $\rho_e = 0.3k\Omega$.

where the central one ($x = 0$) was assumed to lie under the electrode. The FH model was applied only for that node, whereas, the other nodes were simulated with constant ionic conductance.

Before going on into detailed applications, we will summarize the purpose of calculating the activating function:

At the parts of the fiber where the activating function f is positive the fiber is stimulated. But in the areas where $f < 0$ the fiber voltages becomes negative.

The activating function for (small) bipolar or multichannel electrodes will be found by superimposing the influences of single electrodes.

$$f = \frac{\rho_e}{4\pi} \sum_{i=1}^{n} ((x - x_i)^2 + z_i^2)^{-\frac{5}{2}} (2(x - x_i)^2 - z_i^2) \cdot I_{el,i}(t) \quad (7.10)$$

x_i, z_i are the coordinates, and $I_{el,i}$ is the current of the i-th electrode.

Note, that in a homogeneous medium, the influence depends only on the distance of the electrode to the fiber, thereby reducing three-dimensional problems to two dimensions. As an example, we will consider two small spherical electrodes in an homogeneous isotropic medium. It is not important whether or not both electrodes are on the same side of the fiber, on opposite sides, or within a special angle. Important is only the distance between the electrodes and the fiber.

Equation 7.10 is valid for the n 'point' electrode, but it can also be used for small spherical electrodes. Other shapes of electrodes can be approximated by using several point electrodes. For example, if we consider a ring electrode around an axon, we can use this technique. Since only the distance between the electrodes and the axis of the fiber is important, in the special case where the fiber pass the center of a ring electrode perpendicular, the ring electrode can be represented by a single point electrode.

We have seen above, how to calculate the potential in a homogeneous field by neglecting the presence of the fiber itself. In the case of a stimulated nerve bundle, a more accurate model should include differences in current distribution, caused by the presence of the fibers and other inhomogenities. ALTMAN & PLONSEY (1990) have developed a bidomain concept, assuming different conductances in the intracellular and interstitial space. An analytical solution is possible for the stimulation by modeling a dot electrode as point source in a homogeneous field.

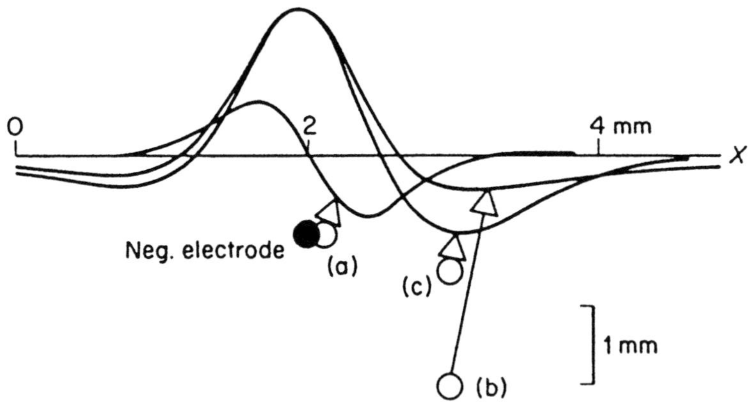

Fig. 7.4 *Activating functions for bipolar electrodes. One electrode is always in the same position, (x=2mm, z=1mm) marked by a full circle. The positions for the other electrodes are 2.2mm/1mm in case (a), 3mm/3mm in (b) and 3mm/1.5mm in (c). In case (b) the influence of the distant electrode is small and the activating function is very similar to that of Fig. 7.2 (C) for monopolar stimulation. In the cases (a) and (c) the axon will fire only in one direction.*

(After Rattay, 1987b)

Stimulation with a dipole

The special case of (7.10) is called the bipolar stimulation when two electrodes are involved with $I_{el,2} = -I_{el,1}$. We assume that the negative electrode is at a fixed position, which is marked with a full circle in Fig. 7.4, and that the position of the positive electrode takes one of the places (a), (b) or (c). For these three cases the activating functions are displayed for a constant I_{el}. In case (b) one pole is relatively far away from the fiber, which is situated on the x-axis in Fig. 7.4. The activating function is similar to that of the monopolar case with symmetric reactions. If the distances of both electrodes to the axon are similar, the activating function has one dominant depolarizing and one dominant hyperpolarizing part (case (a)). The hyperpolarizing part can be strong enough to allow an AP to propagate only in one direction. Thus, if the dipole is supplied with proper alternating superthreshold current, the axon will fire only to one side at the maxima of $I_{el,1}$ and to the other side at the minima. Such one-way stimulations are important for stopping spastic signals or in cochlear implants.

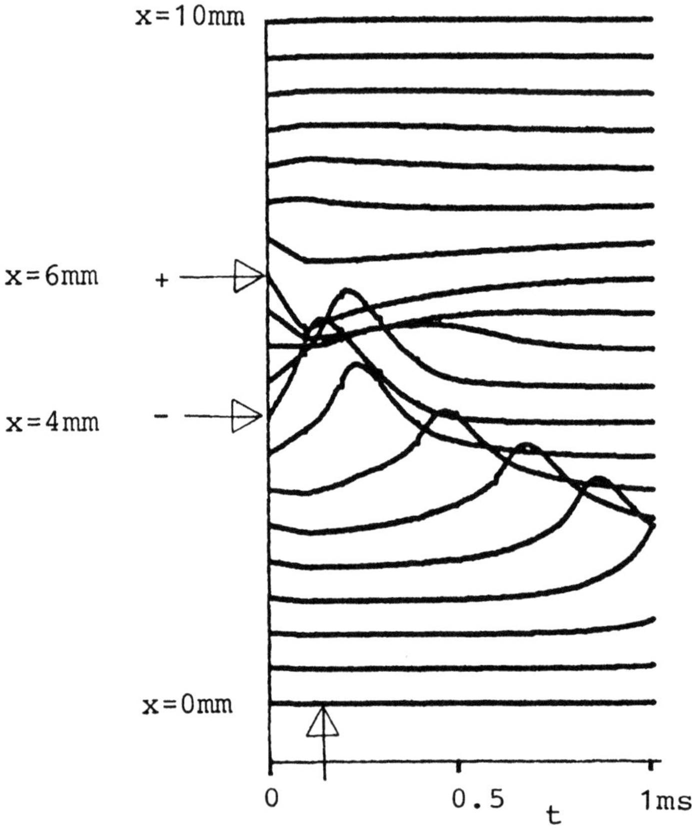

Fig. 7.5 *One-way firing of a dipol electrode. The lines correspond to the points of the axon with a distance of 0.5mm. A cathodic 100 μs current pulse stimulates the axon at the position marked by − → (x=4 mm), and an anodic one of the same magnitude at the position marked by + → (x=6 mm). Both electrodes have the same distance, z=1 mm. The vertical arrow marks the end of the stimulating pulse at t=100 μs. The first linear part of the curves show the depolarizing and the hyperpolarizing influence for the activating function within the stimulus pulse.*

Simulation is with the HH model, according to data of Fig. 6.3, but for extracellular stimulation with electrode currents of -3 mA and 3 mA respectively.

Fig. 7.5 illustrates the one-way firing, for a case like that of Fig. 7.4 (a), in the time domain. At the end of the stimulus impulse (marked by a vertical arrow) membrane voltages are approximately proportionate to the activating function. Depolarization is maximal at the position of the cathode (marked by $- \rightarrow$), hyperpolarization is maximal at the anode (marked by $+ \rightarrow$). The AP tries to propagate to both directions as seen by the lines above and below $- \rightarrow$, which correspond to $x = 4.5mm$ and $x = 3.5mm$. However, the strong hyperpolarization at $x = 6mm$ stops further propagation in the upward direction.

The activating function (7.8) can be approximated by (7.6), i.e., we can approximate the second differential quotient by the second difference quotient. This is especially useful for numerical investigations, when the fiber is segmented with Δx. In the case of inhomogeneous media, with different structures enclosed, or for complicated shapes of electrodes, we will only calculate the extracellular voltage at these points of support.

We will find the same situation when myelinated fiber reactions should be simulated. In this case segmentation length Δx is given by the internodal distance, and we will use the difference quotient (7.6). Nevertheless, for a general analysis of myelinated fibers, it is often useful to approximate (7.6) by (7.8) because we can find results, like those described above for the bipolar electrode, easier with the help of the continuous activating function (7.8). Essential differences will only occur if electrodes are very close to the fiber, i.e., the nodal distance differs distinctly from the axial distance ($z < \Delta x$).

Higher stimuli are necessary to activate the myelinated axons. This is especially apparent if the electrode lies between two nodes of Ranvier. In fact, higher stimuli are necessary compared to the case when it is just above a node. We will discuss this phenomenon in the next chapter.

Excitation under surface electrodes

Although implanted electrodes are more effective, surface electrodes are often used for functional electrostimulation because no surgery is necessary. Unfortunately, the calculation of the extracellular potential is not as simple as in the case of the small implanted electrodes which were discussed above. For several applications we should calculate the extracellular potential using,

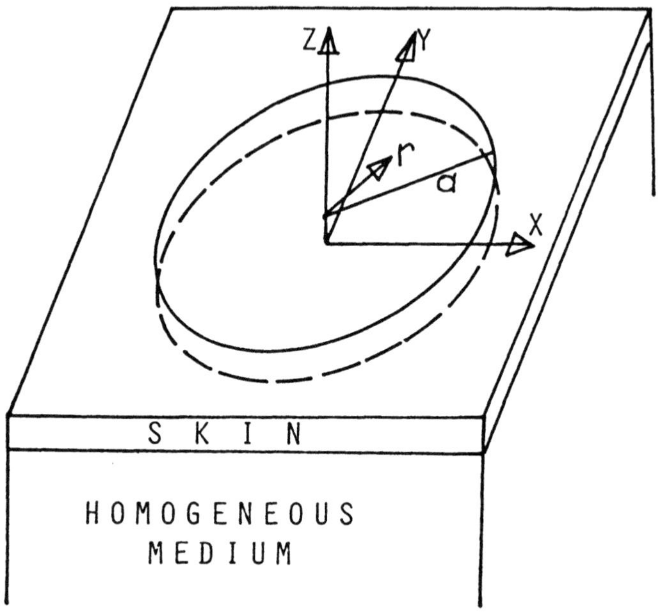

Fig. 7.6 *Geometry of an idealized surface electrode. A circular electrode of radius a is symbolized by the full curve. The semi-infinite homogeneous conductive medium under the electrode is separated by the skin. We will neglect the current spread within the skin and assume that a constant voltage* V_0 *exists within the circle (broken line) at that part of the surface of the homogeneous medium which lies under the electrode.*

(From Rattay, 1988a)

e.g., finite difference of finite element techniques. We will not go into details here, but interested readers will find more information, e.g., in ZIENKIEWICZ, 1971.

The effects of surface electrodes are greatest close to the electrodes. To analyze reactions of fibers in the vicinity of the electrodes we need to simplify the geometry. In order to avoid complications with the non-uniform resistance of the skin, we neglect the current spread within the thin sheets of the skin. We assume that a constant potential V_0 exists in the surface domain of the conducting medium under the skin, which lies just under the electrode (Fig. 7.6). [As a consequence of the resistance of the skin, V_0 is lower than the potential of the electrode.] Furthermore, we will assume that a semi-infinite homogeneous medium exists under the skin; where the nerve and muscle fibers are imbedded. A ground electrode may be at infinity.

In order to find the potential V in the conductive medium we have to solve Laplace's equation

$$\Delta V = 0 \tag{7.11}$$

For electrical potentials with axial symmetry, the solution is of the form (cf. JACKSON 1962; SNEDDON 1966; WILEY & WEBSTER 1982.)

$$V(r,z) = \int_0^\infty A(k) \cdot e^{k|z|} J_0(kr) dk \tag{7.12}$$

with the Bessel function J_0 and suitable $A(k)$ to satisfy the boundary conditions. In our case they are

$$\begin{aligned} &V = V_0 && \text{for } z = 0,\ r \leq a \\ &\frac{\partial V}{\partial z} = 0 && \text{for } z = 0,\ r > a \\ &\quad\text{and} \\ &V \to 0 && \text{for } r \to \infty,\ z \to -\infty \end{aligned} \tag{7.13}$$

Equation (7.11), together with the boundary values of (7.12), has the following analytic solution

$$V(r,0) = V_0 \qquad z = 0,\ r \leq a \tag{7.14a}$$

in the surface domain under the electrode and

$$V(r,z) = \frac{2V_0}{\pi} \cdot \arcsin \frac{2a}{\sqrt{(r-a)^2 + z^2} + \sqrt{(r+a)^2 + z^2}} \tag{7.14b}$$

everywhere else.

Having in mind that the solution is symmetric to rotation we will simplify the geometry by selecting a special direction to define the x axis (Fig. 7.6).

Fig. 7.7 (A) shows the potential V at x lines in two planes parallel to the surface, at depths of $z = -a/10$ and $z = -a/2$. The corresponding activating functions for excitable fibers, which are positioned at those x lines, are plotted in Fig. 7.7 (B). Since the voltage drops quickly at the edge of the electrode, one can really

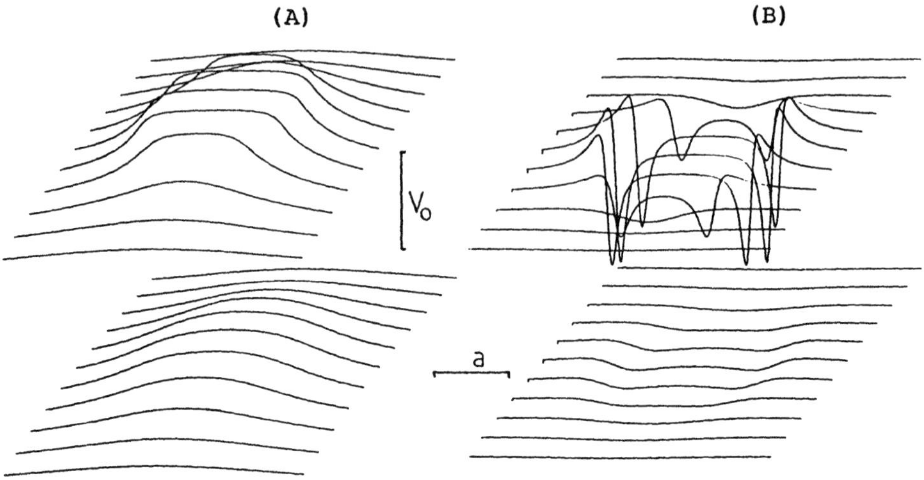

Fig. 7.7 Potentials and activating functions below a surface electrode of radius a. (A) Potential distribution at x-lines as part of V_0*, at the depth of* $z = -a/10$ *(upper diagram) and at* $z = -a/2$ *(lower diagram). (B) Activating functions at the x-lines as in (A). The result is independent from the conductivity. Assumptions are like those in Fig. 7.6; x and y vary from -2a to 2a.*

(From Rattay, 1988a)

see the shape of the electrode, especially in the upper traces of Fig. 7.7 (A) and (B). At a constant depth, the strongest reactions are at the lines which intersect the z axis (y=0).

Fig. 7.8 shows the extracellular potential V and the activation functions $\partial^2 V/\partial x^2$, scaled by V_0 for $y = 0$, at the depths $z = 0$, $z = a/10$, $z = 2a/10$, ... , $z = a$. Because the activating function shrinks with depth, higher voltages are necessary at the electrode to stimulate the deep fibers. The threshold voltages are listed in Table 7.1 for cathodal (V_0^+) and anodal (V_0-) stimulation, for a disc electrode with radius $a = 1cm$. It is seen that for fibers lying parallel to the surface, in the depth z under the electrode, the voltage-distance relation is fairly linear.

As mentioned above, spikes will be generated in areas where the activating function f is positive. If $V_0 > 0$, their (two symmetrical) places of origin will be within the intervals $a <| x |< 2a$ where $f > 0$ (Fig. 7.8 (B) and Fig. 7.9 (B)), but for $V_0 < 0$ the activation function changes the sign and the spikes are generated at $| x |< a$. [In Fig. 7.8 (B) the activating function $\partial^2(V/V_0)/\partial x^2$

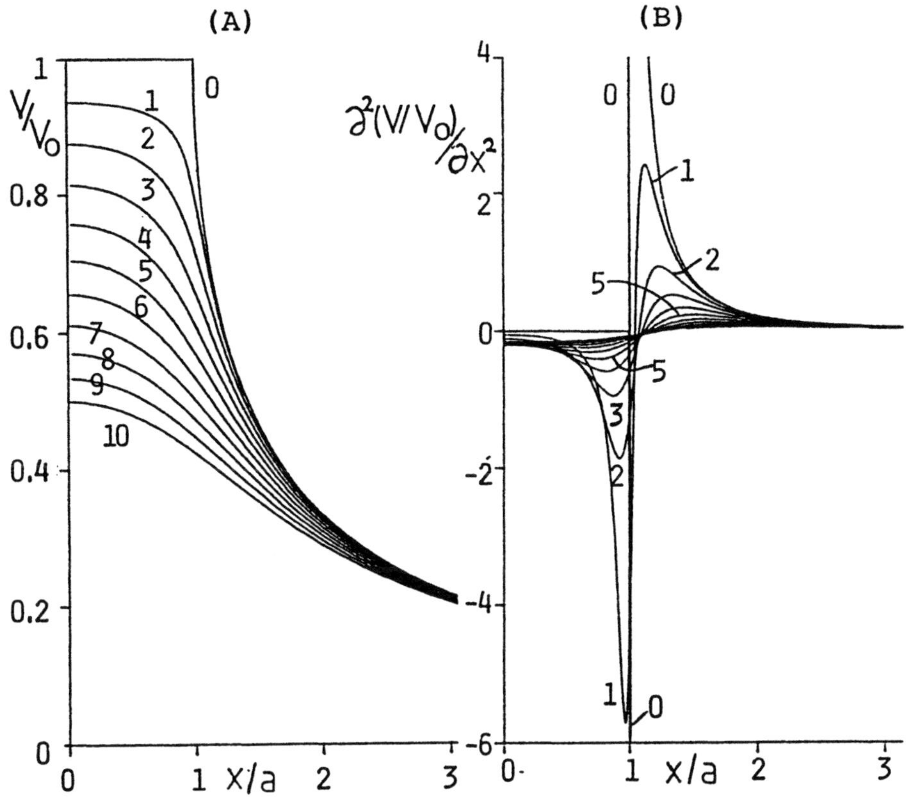

Fig. 7.8 *Normalized extracellular potential V/V_0 and normalized activating functions $\frac{\partial^2(V/V_0)}{\partial x^2}$, as functions of x/a. (A) Normalized potentials for different depths z/a=0, −0.1, −0.2, ... −1. The curves are marked with numbers 0, 1, 2, ... 10 respectively. [The curves 1 and 5 lie in the depth $z=-a/10$ and $z=-a/2$ respectively. These curves are also contained in Fig 7.7 (A) as central curves.] (B) shows the corresponding activating functions. At z=0, the potential (curve 0) is constant under the electrode, but it decreases with a vertical decay at $x=a$. The activating function becomes infinite there. All the curves are symmetrical with respect to x=0.*

(From Rattay, 1988a)

is negative for $|x| < a$.] Fig. 7.9 illustrates those different places of origins of spikes for cathodic and anodic stimulation.

Fig. 7.8 (B) shows that the activating functions have stronger

Table 7.1
Threshold voltages for a surface elctrode
as a function of depth z

	calculated with					
	FH		model	HH		
z	V_0^-	V_0^+	$-\frac{V_0^+}{V_0^-}$	V_0^-	V_0^+	$-\frac{V_0^+}{V_0^-}$
[cm]	[Volt]			[Volt]		
0.1	-0.9	1.8	2	-0.45	0.7	1.5
0.2	-1.8	3.4	1.9	-0.7	0.9	1.3
0.3	-2.7	4.4	1.6	-0.9	1.3	1.4
0.4	-3.8	6.2	1.6	-1.1	1.7	1.5
0.5	-5.2	8.5	1.6	-1.5	2.2	1.5
1	-11.2	29	2.5	-2.8	6	2.1

V_0^- and V_0^+ are threshold voltages for cathodic and anodic stimulation respectively. $-\frac{V_0^+}{V_0^-}$ gives the ratio of anodic to cathodic threshold.

Diameter of the electrode $a = 1cm$. Stimulation was simulated with single square pulses of $100\mu s$ in all the cases. The FH values were calculated with the original standard data but, $T = 37^oC$, $d = 5\mu m$, $\rho_i = 100\Omega.cm$, $\Delta x = 1mm$, and $L = 2.5\mu m$. The HH values were calculated with standard data, but $T = 29^oC$, and $\Delta x = 1mm$.

excursions into the negative domain; therefore, thresholds are smaller for cathodic stimulation than for anodic. This is also demonstrated in Table 7.1. Comparing the columns of the table, we see that the ratio of anodic to cathodic threshold voltages in the HH model is smaller than that for the FH model. We will explain this effect, which is a consequence of different membrane kinetics.

The lowest trace of Fig. 7.9 (A) shows the reaction of a fiber stimulated with a negative pulse. The fiber is at rest before it is stimulated, but after a stimulus pulse of $100\mu s$, V will be always positive. [Arrows in Fig. 7.9 (A).] This means that, in the FH

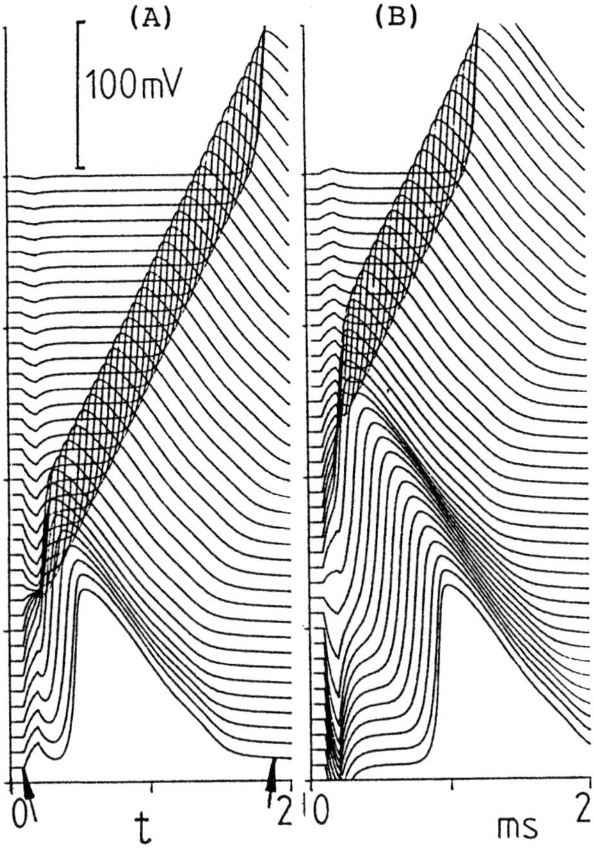

Fig. 7.9 *Fiber reaction to negative square pulse ($V_0 = -6V$) (A) and to a positive pulse ($V_0 = 12V$) (B). Simulation with Frankenhaeuser-Huxley original standard data, but $d = 5\mu m$, and $T = 37°C$. The duration of impulse is $100\mu s$, nodal separation $\Delta x = 1mm$, $L = 2.5\mu m$, $\rho_i = 100\Omega.cm$.*

Every line shows the reaction at a node. The lowest line corresponds to $x = 0$, the next one to $x = 1mm$ etc. until x=40mm. For negative x values, the result is a mirror image of the one shown. Within stimulation time, from t=0.1ms to t=0.2, the conductance is nearly constant because the stimulus signal is just a little above threshold. We can recognize the image of the corresponding activating function, marked by number 5 in Fig. 7.8(B). Arrows mark the voltage before ($V = 0$) and after an action potential ($V > 0$, no hyperpolarization).

(From Rattay, 1988a)

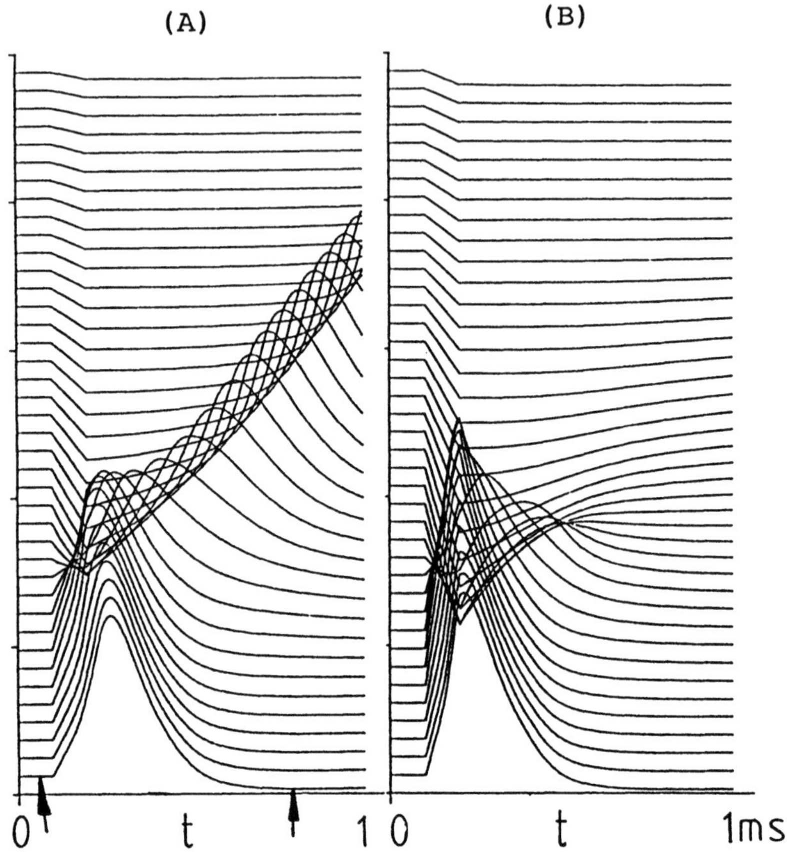

Fig. 7.10 *Reaction of a fiber under a surface electrode, simulated with the HH-model. Computation is based on the data of Table 7.1, in the depth z=–0.5cm and with a stimulus voltage of –5 V in (A) and –10 V in (B) (threshold: V_0^- = 1.5 V). Fiber activity will not propagate in (B) because the spike cannot travel across the strongly hyperpolarized region. Arrows mark membrane voltage before (V=0), and after an action potential (V < 0, hyperpolarization).*

(From Rattay, 1988a)

model, the 'natural' shape of a spike has no or nearly no hyperpolarized region, which is in contrast to the HH axon. [Arrows in Fig. 7.10 (A); see also Fig. 5.9.] Generally, we can say that the Hodgkin-Huxley membrane is more sensitive to hyperpolarization than the Frankenhaeuser-Huxley membrane.

To explain the "hyperpolarizing effect" we will compare the first line of Table 7.1 with the corresponding activating function marked by 1 for z=a/10 of Fig. 7.8(B). The negative part of the scaled activating function is more than twice as strong as the positive one [Fig. 7.8(B)] and the FH ratio of threshold voltages V_0^+/V_0^- is equal to two. The threshold voltage V_0^+ needed to excite the giant Hodgkin-Huxley axon at the same position ($z = -1$ mm) is 0.7 V, but a -0.35 V pulse is too weak to overcome the hyperpolarized region at the outside of the electrode ($x > a$). By line 1 of Table 7.1 we find that the ratio of $V_0^+/V_0^- = 1.5$.

Fig. 7.10 shows that the blocking effect is also possible with surface electrodes. Strong cathodic stimuli of about 8 times threshold can block the propagation of an AP. This phenomenon is found by simulation with the HH model, but it is not seen with the original FH-model (RATTAY, 1988a).

WILEY & WEBSTER (1982) have shown that surface electrodes, as described above, produce high current densities at the edges. To avoid burning in these areas, several methods are in use. There exist electrodes, which consist of concentric rings, that allow stimulating voltages to drop against the outside. Another possibility is the use of electrodes with low conductance at the edges. In that case V_0 is not constant, and the small radii of curvature will disappear in Fig. 7.8 (A) for small depths.

8. CURRENT-DISTANCE RELATIONS FOR MONOPOLAR ELECTRODES AND FOR RING ELECTRODES

Comparison between simulated and experimental data

For many applications using implanted electrodes, it is important to know how many fibers can be activated by a given current signal. A simple rule is: distant fibers need stronger currents in order to be stimulated, and also the number of firing fibers increases with the strength of stimulation signal. Fig. 8.1 illustrates the current-distance relation from several experiments which were gathered by RANCK (1975). Because there are differences in the original experiments, other durations of current impulses are considered by a correcting factor. The dashed line shows the calculated relation with the assumption that the electrode lies just above a node. Simulation was done with equations 7.2, 7.5 and the FRANKENHAEUSER-HUXLEY model.

BEMENT & RANCK (1969) reported that the slopes of the current-distance lines are inversely related to the conduction velocity of the fibers. JANKOWSKA & ROBERTS (1972), as well as, JANKOWSKA & SMITH (1973) reported a minimum threshold of $0.1\mu A$ resp. $0.2\mu A$ for myelinated fibers of the spinal cord when the electrode is placed very close to a node. ROBERTS & SMITH (1973) suggest that the threshold current is approximately proportional to the square of the distance between the node and the electrode, for distances less than $150\mu m$. However, their experimental arrangements do not guarantee exact measurements of the nodal position. We have obtained other results by simulation, which is discussed below.

Experiments with $0.1ms$ superthreshold cathodic stimulation show that currents greater than 8 times threshold cannot produce a propagating action potential in the case of myelinated fibers (ROBERTS & SMITH, 1973). A similar effect was found for small unmyelinated fibers (KATZ & MILEDI, 1965). A special fiber will be stimulated, if a cathodic electrode current is above threshold, but below the value where the superthreshold blocking phenomenon occurs. Again, we assume a homogeneous extracellular medium, where the fiber and the monopolar electrode is embedded. All fibers should be at rest before being stimulated.

In order to distinguish whether a fiber of diameter d is stimulated or not we can use two spheres as shown in Fig. 8.2. All fibers which pass through the outer sphere but do not enter the inner sphere are stimulated (case 2). RANCK supposes that for smaller fiber diameters both spheres are smaller and therefore it is possible to stimulate axons with smaller diameters even when they are nearer. An explanation of such phenomena is possible with the help of the activating function and simulations with equations 7.5 and 7.9.

Unmyelinated fibers

Figure 8.3 shows the computed current-distance relation for an unmyelinated fiber when stimulated with a current pulse of $100\mu s$. Calculation was done according to (7.5) with $L = \Delta x = \frac{z}{2}$ (L: length of the active membrane part of one segment, Δx: segmentation length, z: distance between electrode and axon). We will get exactly the same results for a fiber diameter four times greater by doubling Δx, z and electrode current I_{el}. Therefore, the doubled velocity is caused by multiplying the diameter d with a factor of 4. The outer scales of Fig. 8.3 belong to that four-fold diameter. We can apply the diagram of Fig. 8.3 to arbitrary fiber diameters (measuered in μm) by multiplying the values for the scales of the distance and the current with $\sqrt{\frac{d}{9.6}}$.

The vertical lines (a) and (b) of Fig. 8.3 help to find the distances within which fibers are activated for $I_{el} = -4mA$ and $d_1 = 9.6\mu m$ resp. $d_2 = 38.4\mu m$. The spheres, within which AP's are produced, are shown in the inserted part of Fig. 8.3. Several physiologists hypothesize a linear dependence exists in the current-distance relation, but simulation shows that a linear approximation holds only for small ranges. Assuming linear limits (and $I_{el} = 0$ at $z = 0$) and regarding the lines (a) and (b) we can conclude that the threshold is independent from the diameter. This is in conflict with experimental experience.

The following results for cathodic monopolar stimulation which are illustrated in Fig. 8.3, are also confirmed by the experiments as discussed above:

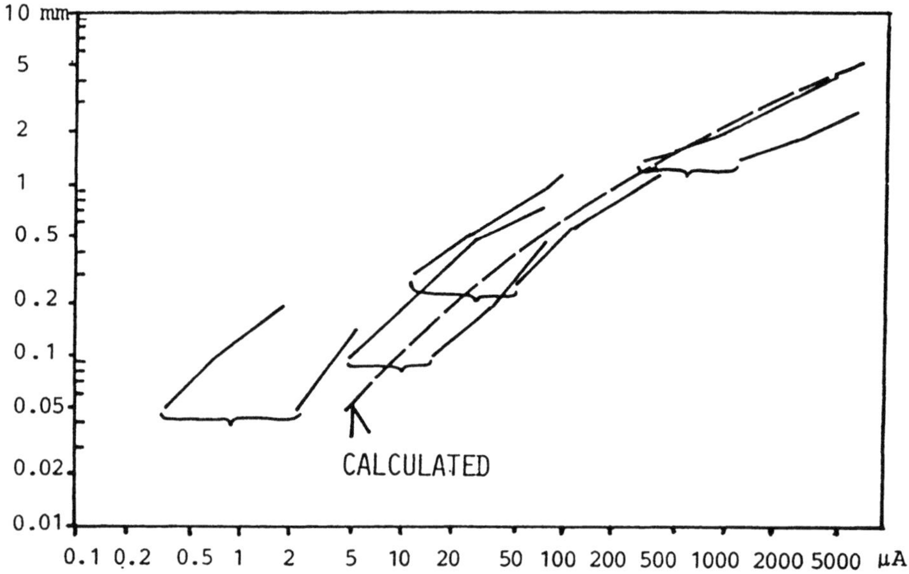

Fig. 8.1 *Current-distance relation for monopolar cathodic stimulation of myelinated fibers. Experimental data of different authors were gathered by Ranck (1969). Their threshold currents are distributed within the left and right lines. Thresholds of experiments of the same investigators are marked with brackets. The dashed line gives the computed current-distance relation with the orignal FH model.*

For the calculations it was assumed that the electrode lies just above a node. Note that the slopes in the diagram are flatter for greater distances. The dashed line starts under an angle of 45°. *This means that the electrode threshold current is proportionate to the distance of the electrode to a node. For distances* z *in the order of* 1*mm we obtain an approximately quadratic relation (doubling the distance needs 4 times the current). For greater distances the slope of the current-distance curve is even flatter and we theoretically obtain an increase of the electrode current proportionate to the third power of the distance.*

Simulation is with Frankenhaeuser-Huxley standard data, but $d = 10\mu m$, $T = 37^oC$, $\rho_i = 100\Omega.cm$, $\rho_e = 300\Omega.cm$. *Stimulation with a* $200\mu s$ *current pulse, which is also the basis for the diagram of the experiments.*

(After Ranck, 1975 and Rattay, 1988a)

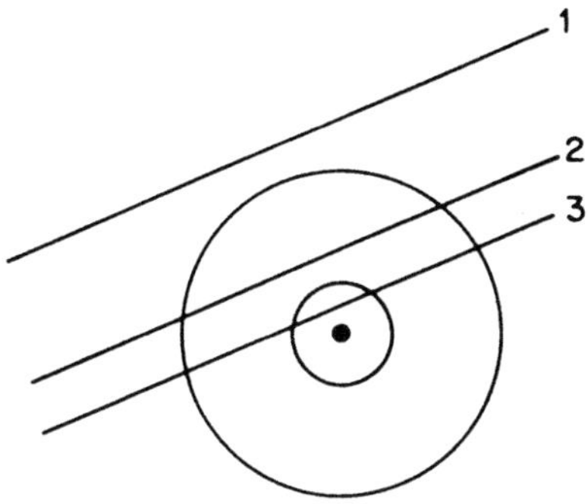

Fig. 8.2 *Cathodic stimulation produces three regions around the electrode, which is marked by the small full circle. Axon 1 lies in the subthreshold region, axon 3 in the blocking region, where spikes cannot propagate, and axon 2 is firing. The diameters of the spheres depend on fiber diameter and on stimulus strength and duration.*

i) Stimulation reaches thicker fibers in areas farther away from the electrode.
ii) The lower limit of electrode distance (blocking phenomenon) also rises with the diameter.
iii) Within a small area, defined by two spheres, small fibers with diameter d_1 are stimulated by a given pulse. But fibers with a greater diameter d_2, are not stimulated.

Following the horizontal line (c) of Fig. 8.3, we can see that currents stronger than about 8 times the threshold will lead to a cessation of cathodic stimulation. This also follows from the experiments of ROBERTS & SMITH (1973) and KATZ & MILEDI (1965) as mentioned above.

A strong positive pulse will also stimulate the fiber (Fig. 8.3). No AP is generated at $x = 0$ (as in the case of cathodic monopolar stimulation), but in the outside areas two AP's are generated and propagate outwardly. These are located at two symmetrical positions, according to the polarizing parts of Fig. 7.2(B).

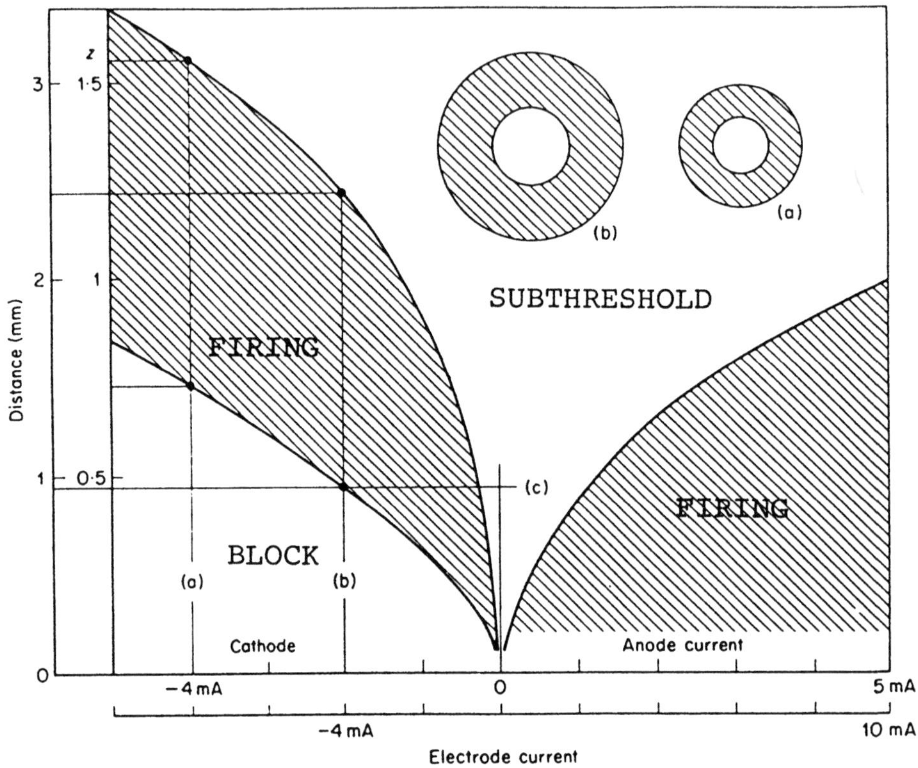

Fig. 8.3 Current – distance relation for unmyelinated fibers. Stimulation with a monopolar spherical electrode produces firing in the shaded area. The inner scales belong to a fiber diameter of 9.6μm, the outer ones to 38.4μm. Line (a) shows the lower and the upper limit for cathodic stimulation with an electrode current of 4mA. The same signal reaches fibers more distant from the electrode if their diameters are greater, as shown by line (b) for d = 38.4μm. The 'firing' regions are limited by spheres, as illustrated by the inserts for these cases. Line (c) shows the lower and the upper boundary for the electrode current. This allows us to stimulate a fiber at a given distance from the electrode.

Computation was done with Hodgkin-Huxley standard data, but for changed diameters, $T = 27°C$, and $\rho_e = 300\Omega cm$. A square pulse of 100μs was used as stimulus.

(After Rattay, 1987b)

Myelinated fibers

There are two differences between the simulation of myelinated and unmyelinated fibers. Firstly, the ionic currents are computed with other equations (e.g., FH-model instead of HH-model). The most striking difference is in the gating mechanism, which was already seen in the results of the space-clamp experiments: HH axons showed slow subthreshold responses (flat part in Fig. 5.10 (B)) even for high temperatures which cause short AP's. The FH-model demonstrated that, in a myelinated axon, the membrane voltage droped suddenly against the resting value when stimulated with a subthreshold stimulus impulse (Fig. 5.10 (A)). The same effect appeared in the model of Box 5.3 (Fig. 5.11). Those differences in membrane behavior may cause distinct blocking reactions.

The second difference is caused by the relative long internodal distance in the case of myelinated fibers. The firing behavior depends not only on the distance z between the electrode and the axon, but also on the x-distance to a node. Stimulation is easiest when the electrode is above a node ($x_{el} = 0$), but stronger currents are necessary when it lies between two nodes ($x_{el} = \Delta x/2$) (ROBERTS & SMITH 1973; RATTAY 1987b). The difference between nodal and internodal stimulations becomes dominant when the electrode is very close to the fiber.

Figure 8.4 shows the cathodic threshold currents, for different positions of the stimulating electrode, on a grid with a mesh size of 0.1mm. The full circles mark two nodes and the values on the first line belong to the theoretical distance $z = 0$. As seen from the figure, I_{el} looks roughly linear on z for $0.2mm < z < 0.6mm$. This is in accord with past experiments (ROBERT & SMITH 1973; RANCK 1975). It is interesting that moving the electrode from the node in the x-direction, along the axis of the axon, shows approximately the same threshold currents as moving the electrode from the node in the z-direction (Fig. 8.4), as long as, the distance of the node to the electrode is 30% below that of the internodal distance. This is due to the dominant influence of the very close node. We can see from Fig. 8.4, that **for small distances the threshold current depends mainly on the distance to the node - not on the distance to the axon.**

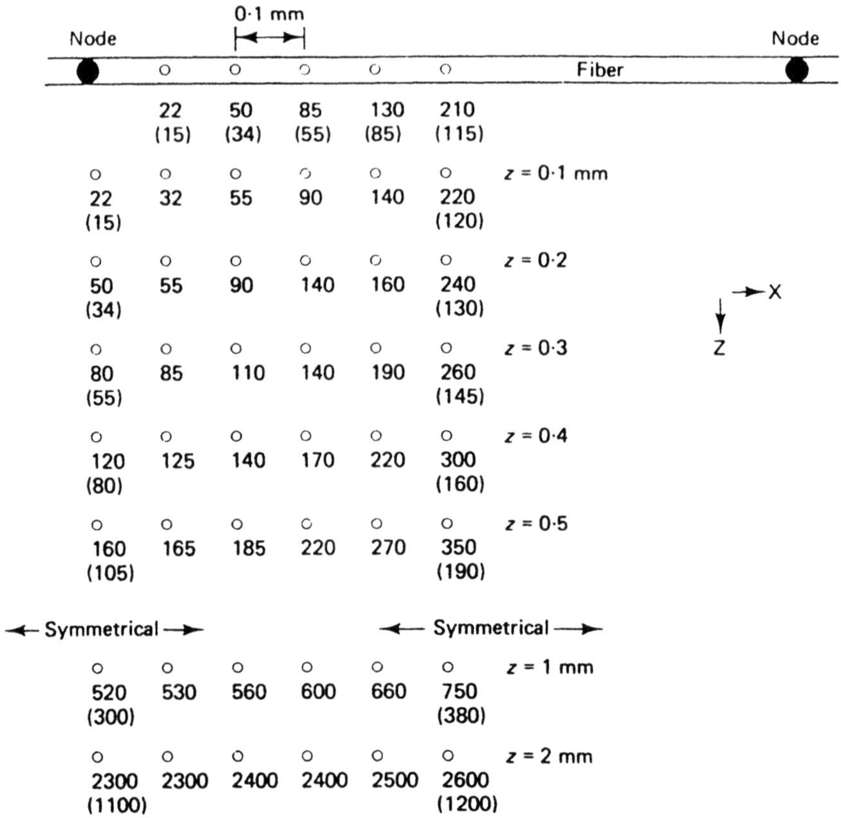

Fig. 8.4 *Cathodic stimulation of a myelinated fiber. The electrode is moved on a grid, with a mesh size of 0.1mm, in the area between nodes of Ranvier. The positions of the electrode are marked with circles, the numbers give the threshold currents in μA for a 0.1ms pulse and for a diameter of $2.4\mu m$. (In brackets: threshold for $d = 4.8\mu m$, but for the same internodal distance of 1mm). The current-distance relations shows two symmetries: above the node (marked with a full circle) and in the middle, between the two nodes. Differences between the threshold values are high for a fixed z close to the fiber, but the thresholds are nearly independent from x for $z > 2mm$.*

Computation was done with original Frankenhaeuser standard data, $\rho_i = 100\Omega.cm$, $\rho_e = 300\Omega.cm$.

(After Rattay, 1987b)

Table 8.1
Stimulus strength (f_n) at the nodes near to the electrode as a function of z

(a) Electrode is above a node: $x_{el} = 0$

z [cm]	node 0 $x = 0$	node ±1 $x = \pm\Delta x$	node $\pm$ 2 $x = \pm2\Delta x$	node $\pm$ 3 $x = \pm3\Delta x$
0.00	$-\infty$	∞	0.07958	0.01989
0.01	-4.29956	2.03145	0.07865	0.01982
0.02	-1.91913	0.84424	0.07595	0.01958
0.03	-1.13422	0.45649	0.07176	0.01920
0.04	-0.75035	0.27056	0.06644	0.01868
0.05	-0.52787	0.16621	0.06042	0.01804
0.10	-0.13985	0.00788	0.03077	0.01368
0.20	-0.02520	-0.00976	0.00417	0.00536

(b) Electrode is between two nodes: $x_{el} = \frac{\Delta x}{2}$

z	$x = \Delta x$ $x = 0$	$x = 2\Delta x$ $x = -\Delta x$	$x = 3\Delta x$ $x = -2\Delta x$
0.00	-0.31831	0.25456	0.03638
0.01	-0.30939	0.24600	0.03615
0.02	-0.28556	0.22299	0.03548
0.03	-0.25336	0.19211	0.03440
0.04	-0.21906	0.15957	0.03296
0.05	-0.18663	0.12928	0.03123
0.10	-0.08110	0.03734	0.02068
0.20	-0.02031	0.00062	0.00558

Listed are stimulus coefficients $\Delta^2 V_e/\Delta x^2$ in mV/cm^2 for $\Delta x = 1mm$, $\rho_e = 0.3k\Omega$ and $I_{el} = 1\mu A$. The extracellular potential was calculated by (7.2).

The bracketed values in Fig. 8.4 are obtained when $d/\rho_i L$ of (7.5) is doubled (e.g., d is doubled but all other parameters are unchanged). Now let us double d and Δx without changing the other parameters*. Here we will have to double the original values for the electrode currents I_{el} of Fig. 8.4 and the new values will correspond to the doubled grid size.

As explained above, the activity of a myelinated fiber is governed by the second difference quotient of V_e, which is related to the activating function for myelinated fibers (7.6), (7.8). The subsequent effects on myelinated and unmyelinated fibers will be similar for z greater than the internodal length.

Table 8.1 shows the 'activating' effects at the nodes near the electrode, namely the second difference quotients of V_e for an electrode current of $1\mu A$. The electrodes are positioned at distances z, as in Fig. 8.4, but only above a node [case (a) for $x_{el} = 0$] and just between the two nodes [case (b) for $x_{el} = \Delta x/2$]. In these cases, the reaction of the fiber will be absolutely symmetrical at both sides of the fiber; symmetry is lost in other cases. The dominant negative coefficients, in Table 8.1, account for the fact that cathodic stimulation requires less threshold currents than anodic stimulation. In case (b), the coefficients of the dominant influence are smaller than in case (a), especially for a small z. This explains the different thresholds.

By inspection of Table 8.1 it is seen that only the first columns have dominant influence on the excitation behavior, especially for small z, which means that only the nodes close to the electrode are involved. If we are interested only in the question: "Is threshold reached in a special fiber ?", we can neglect the nonlinear membrane conductance at the outside segments. The complicated equations for the ionic currents (Chapter 5) are used at few nodes, whereas, at the outside nodes i_{ionic} of eqn. 7.5 is assumed to be proportionate to V_n. McNeal (1976), and after him several other investigators used this reduced system in order to save computation time. In same cases with high computational effort it seems appropriate to assume constant membrane conductances in all nodes (Sweeney *et al.*, 1990).

Cathodic block and stimulation with anodic currents are also known at myelinated fibers. Fig. 8.5 shows the threshold currents

---

* This seems to be the physiological realistic relation between fiber diameter and internodal length. Usually, we have to assume that $\Delta x \sim 100 \cdot D$ and $d \sim 0.7D$

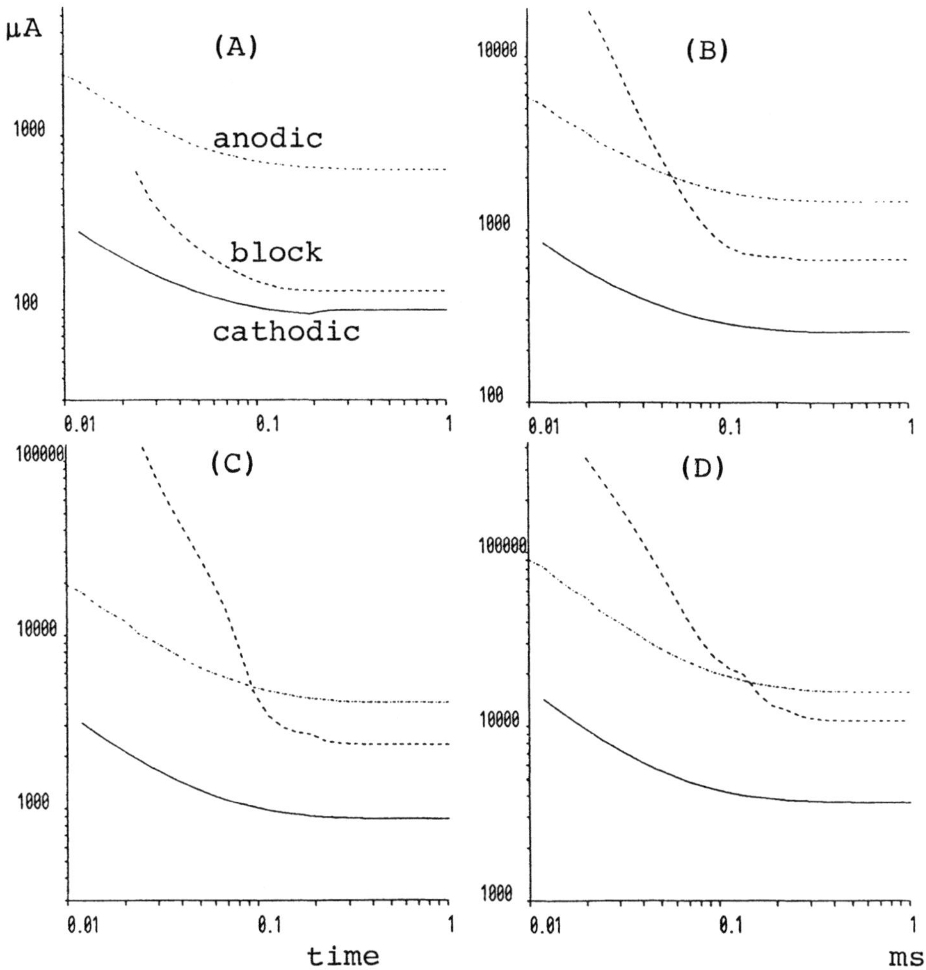

Fig. 8.5 Minimum currents to produce cathodic and anodic excitation, and cathodic block as functions of impulse durations by stimulation with a monopolar electrode. The pulses consist of a plateau (pulse duration) and a linear falling phase of the same duration. The four cases are for different distances between electrode and axon: (A) 0.5 mm, (B) 1 mm, (C) 2 mm, (D) 4 mm. For very short pulse durations the model do not predict blockade. Furthermore, for short pulses the membrane voltage becomes very high. This would cause a demage of the membrane.

The strength-duration relations are calculated with the SE model for 37°C, and a fiber diameter of 10 μm. It is assumed, that the electrode is just above a node.

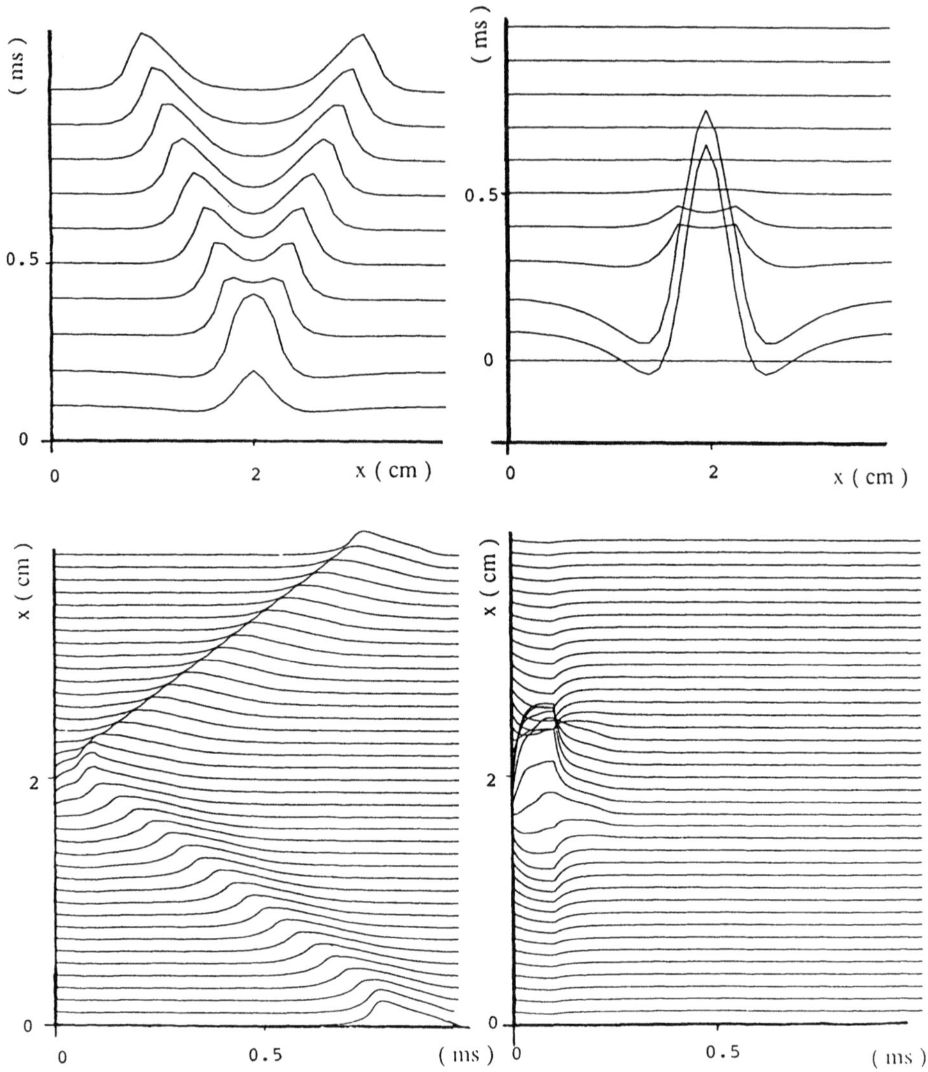

Fig. 8.6 Cathodic excitation (left) and cathodic block (right). Snapshots of membrane voltage show the symmetric reactions of the fiber (above) when stimulated with a monopolar electrode at x=2 cm, 4mm from the axon. Time evolution in the nodes (below) for the same situation.

Simulation with SE model; fiber diameter: 10 μm, pulse duration: 100 μs, stimulus strength: 10 mA for excitation and 80 mA for block.

for cathodic and anodic excitation as well as for cathodic block as functions of pulse duration, simulated for electrode distances from 0.5 *mm* to 4 *mm*. The curves for excitation in Fig. 8.5 are almost parallel, indicating that anodic threshold is about 6 times greater than the cathodic threshold. [Variance is found from 4.5 to a factor 8.]

The relation between the thresholds for excitation and block is more complicated: it depends on the distance between the electrode and the fiber; for pulse durations shorter than 100 μs the block currents become very strong. Therefore, long pulse durations are preferred in blockade experiments. Both, the experiment and the simulation show that strong cathodic currents seem to block the AP, but finally they will fire when the stimulating impulse is switched off. This will not occur if the plateau phase is followed by a falling phase, which may be linear or exponential.In Fig. 8.5 the stimulating signals have a linear falling phase as long as the plateau.

Fig. 8.6 shows the membrane voltage along the fiber for two cases: excitation and block by negative electrode currents. All the results, shown in Fig. 8.5 and 8.6, have been calculated with the SE model, but similar results will be obtained by using the other models for warm-blooded temperatures. However, it is important to notice that **the FH model fails to predict cathodic block when used with the original data.**

Monopolar electrodes versus ring electrodes

With the McNeal method, described above, Veltink *et al.* (1988) calculated the excitation of nerves by using probability distributions of a number of excitation parameters for two cases. These are easy to compute because they are symmetric to rotary, namely the stimulation with a central intrafascicular point electrode and an extraneural ring electrode. They assumed that there were different conductances within the fascicle, in the connective tissues surrounding the fascicle and in the extraneural medium. Furthermore, the inside medium of the fascicle was assumed to be anisotropic with different conductances in the axial and in the radial direction.

Figure 8.7 shows the expectation of stimulated fiber populations as a function of stimulus current. Unfortunately, all the

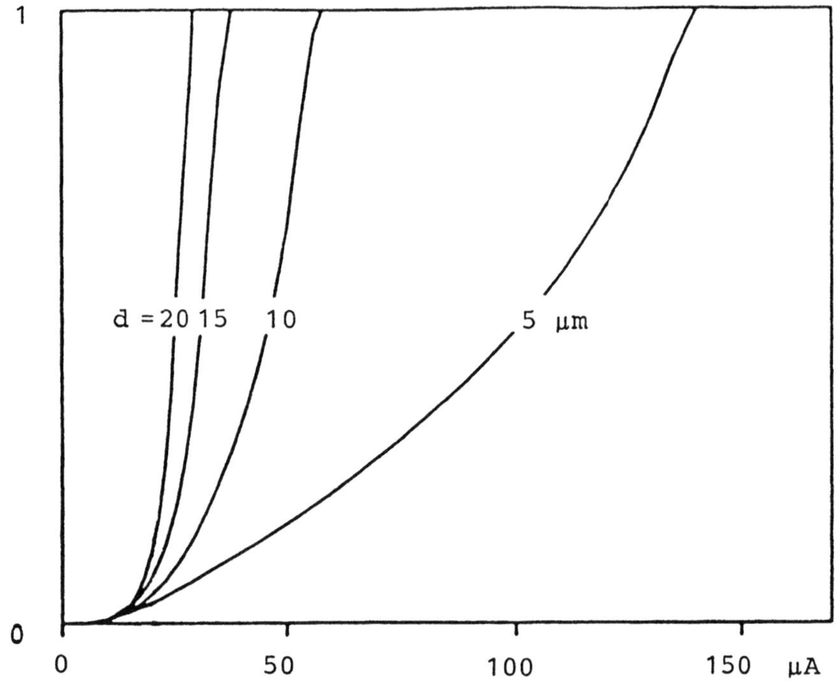

Fig. 8.7 Probability of excitation as a function of stimulus strength for different fiber diameters, when stimulated with a central point electrode. It is assumed that the fibers have internodal lengths Δx which depend on diameter, but the nodes are distributed uniformly respectively to the electrode. Fibers of a special diameter are assumed to also be distributed uniformly in the fascicle. Stronger (cathodic) signals stimulate more fibers and, for a special strength, all fibers of diameter d are firing. For example, with a current amplitude of $30\mu A$ all fibers with $d > 20\mu m$ are stimulated, but only about 15% of the fibers with $d = 10\mu m$ are stimulated. The fibers which will not fire, because of cathodic block, are neglected in the diagram.

Simulation was done with the original FH model at the 3 nodes closest to the electrode and with a constant membrane conductance at the next 8 nodes. The medium, where the axons are embedded, is described by different specific resistances: The intrafascicular resistance in the radial direction is $1.25k\Omega.cm$, and in the axial direction $0.2k\Omega.cm$ (After Geddes and Baker, 1967). Diameter of the fascicle: 1mm. This inner cylinder is surrounded by 0.5mm of connective (isotropic) medium with a $\rho = 1k\Omega.cm$. It was assumed that the nerve was isolated against the outside by the use of a cuff; therefore, the extracellular resistance was $100k\Omega.cm$. Pulse duration: $60\mu s$.

(After Veltink et al, 1988)

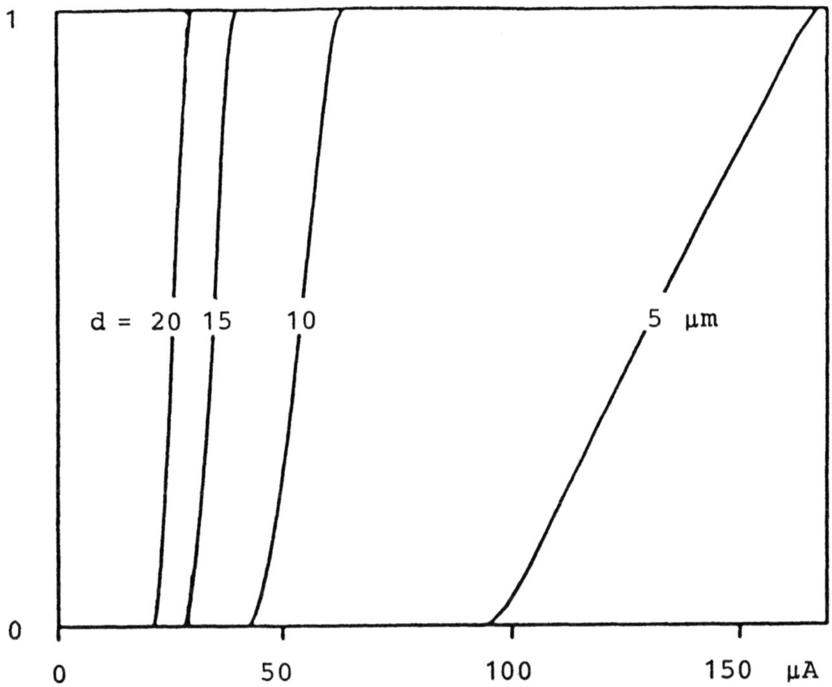

Fig. 8.8 Probability of excitation, as a function of stimulus strength, for different fiber diameters, when stimulated with a ring electrode. An extraneural ring electrode with a diameter of 2mm was used to stimulate the fibers in a nerve. All assumptions as in Fig. 8.7.

(After Veltink et al, 1988)

thick axons are stimulated before only half of the $5\mu m$ fibers are activated. This gives bad recruitment characteristics for motor nerve stimulation, because the thick axons stimulate bigger parts in the muscle and the control of power is within a relative small range of the stimulating current (perhaps between 10 and $40\mu A$ in our example).

The recruitment order is even worse in the case of ring electrodes (Fig. 8.8), because there are no fibers with nodes close enough to the electrode, to obtain the very low threshold currents, as in the last case. As a consequence, the curves in Fig. 8.8 have no flat parts in the beginning, like those seen in Fig. 8.7.

The population of fibers, which are firing when a special stimulus is applied, can be estimated by the activating function. Ring electrodes, as well as, other types of implanted electrodes can

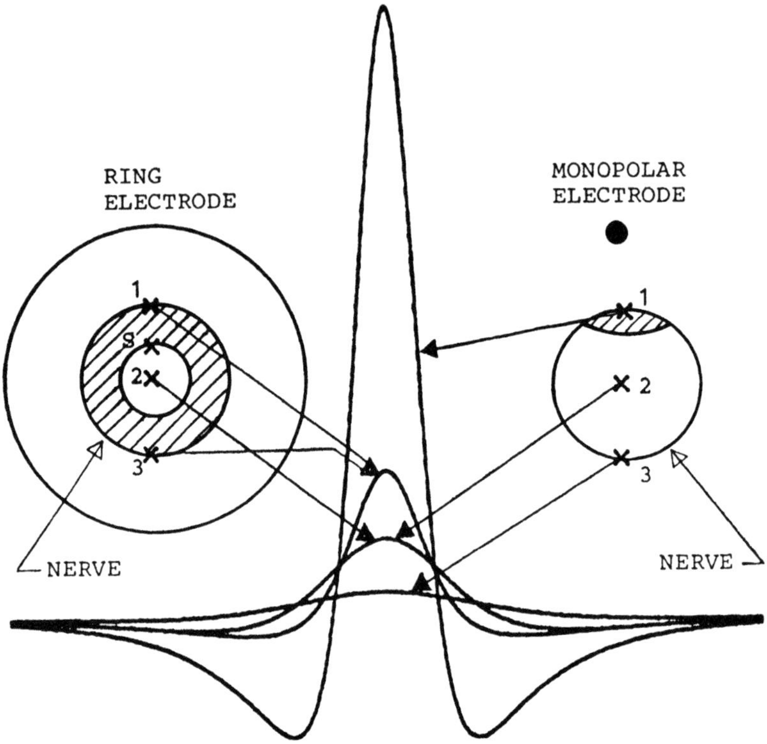

Fig. 8.9 Activating functions for ring electrodes and for monopolar electrodes. The activating functions for three different fibers in a nerve are marked by arrows. In the case of stimulation with a ring electrode the situation is symmetric to rotary; therefore, case 1 and case 3 give the same result. The central fiber 2 is hardest to stimulate, and because the maximum of the activating function for this position is not considerably smaller than for fibers at the edge (1, 3), the ring electrodes have bad recruitment characteristics. Monopolar electrodes show much more variations in the amplitude of f. If the same current signal is applied at both types of electrodes we obtain exactly the same f for the central fiber 2. In the case of a ring electrode the amplitude of f at a point S (in the middle of position 1 and 2) is 61% of point 1. If we assume that I_{el} *brings a fiber at S to threshold, all fibers with the same or greater diameter lying in the shaded area (which is 75% of the total nerve area) will fire, too. In the case of a monopolar electrode the same situation* ($f_{max,S} = 0.61 \cdot f_{max,1}$) *excites a much smaller population of fibers (shaded area). The diameter of the ring electrode was assumed to be twice the diameter of the nerve, and the same as the distance of the monopolar electrode to the central fiber 2.*

(After Rattay, 1989)

be approximated by a number of single electrodes. By means of (7.10) the corresponding activating functions are calculable. Fig. 8.9 shows a small difference in the activating functions, obtained for one of the most excitable points at the edge of the nerve, and for the point in the center, which is most difficult to excite. The small variation of f is the reason for the bad recruitment order obtained from experiments as well as from other calculations (Fig. 8.8). Much better results are obtained by the theory for the stimulation with a monopolar electrode, which is also demonstrated by Fig. 8.9.

Dipolar stimulation

As another example, Fig. 8.10 shows an estimation of the population of fibers stimulated from a dipole, which lies in a plane perpendicular to a nerve. We assume that all fibers have the same diameter and that they are stimulated by inverse current impulses at both electrodes ($I_{el,1}(t) = -I_{el,2}(t)$). For a special current strength, all fibers will be stimulated in the shaded area.

If we also consider fibers with other diameters, or if we allow statistical distribution of the nodes of Ranvier, the border between excited and non-excited fibers is not sharp and cannot be defined by lines (for evaluations see above). Nevertheless, the activating function gives basic information on the recruitment order for special geometries. In the case of our dipole example, we see that it is not possible to stimulate fibers at the central line marked with 0 and we get a much better recruitment order than for a ring electrode.

The dipole described above is disadvantageous because we cannot activate the central region of the nerve. This is caused by the assumption that the nerve is perpendicular to the dipole. However, we are not concerned with this problem in the general case: If the dipol is not perpendicular to the nerve we find for a fiber which is in the center (as well as for all other axons) a situation as described in Fig. 7.4. In contrast to the symmetric reactions produced by the perpendicular situation, uni-directional firing is possible, now.

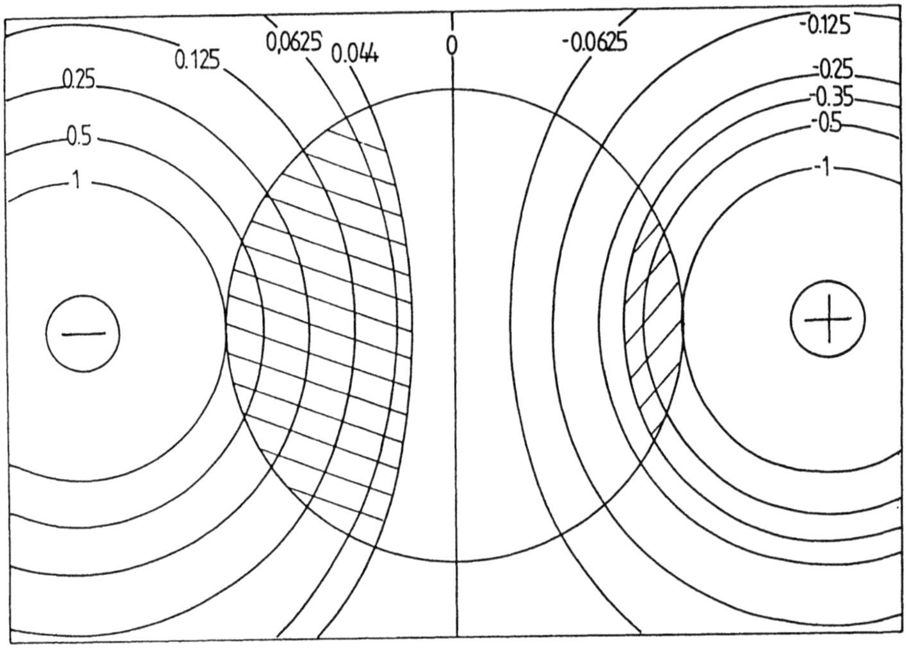

Fig. 8.10 *Nerve stimulated by a dipole ($I_{el,-} = -I_{el,+}$). The central circle marks the boundary of a nerve which is assumed to be perpendicular to the plane of the stimulating dipole. At the lines around the poles, the activating functions have the same maximum values. The numbers give the scaled maxima of f. Because of the broadening of the stimulating part of the axon, (compare Fig. 7.2(d)) we only need an 8 fold current to stimulate the fibers at the line 0.044 - compared with the most excitable fibers at line 1 at the boundary of the nerve. At the anodic part, a smaller population of fibers is stimulated. Simulation with Frankenhaeuser-Huxley standard data, $\rho_e = 300\Omega.cm$.*

(From Rattay and Mayr, 1987)

9. REPETITIVE FIRING AND FIBER REACTIONS TO PERIODIC STIMULI

Stimulation with constant current

Since the shape and the amplitude of the propagating AP does not carry any information, the neural system uses firing sequences in order to transmit information over distances greater than several millimeters. Many nerve cells are able to generate trains of impulses of constant or varying frequencies. Rythmic firing of the pacemaker cells is modulated by the transmitters adrenaline and acetylcholine, but the firing frequencies of axons can also be modulated by sensory stimuli, synaptic potentials, or by electrostimulation. (CONNOR & STEVENS 1971 a,b,c; several further references are found, e.g., in JACK *et al.* 1983).

Repetitive firing can be evoked by applying periodic or constant stimulus signals. First, we will consider constant currents. An example is shown in Fig. 9.1 which was calculated with the HH model (space clamped), thereby, the following typical results were obtained:

i) A low superthreshold current generates an AP only when switched on.
ii) Within a certain range the stimulus signal produces repetitive firing.
iii) Stronger currents produce quasiperiodic solutions of higher frequency, but the voltage does not reach the amplitude of an ordinary AP.

GUTTMAN & BARNHILL (1970) found a good agreement between the experimental data (Fig. 9.2) and the model calculations for different temperatures of the uniformly polarized (space clamped) squid axon membrane. It is seen that the firing frequency varies much more with temperature than with current strength.

Since the amplitude of the quasiperiodic solution of an uniformly polarized membrane is smaller than that of an AP, we can ask the question: Which current strength will produce propagating signals ?

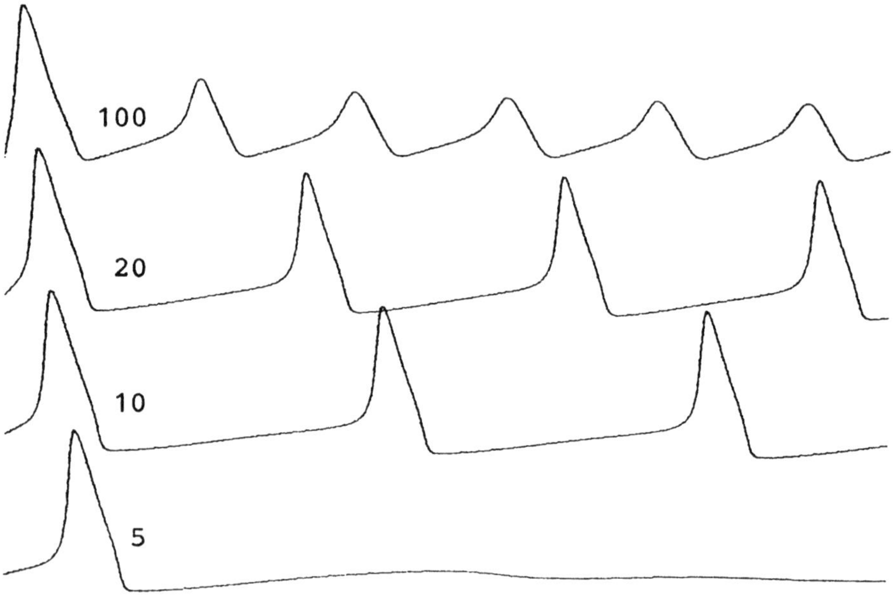

Fig. 9.1 *Repetitive firing of the space clamped HH-model. Constant currents, which are stronger than $2.3\mu A/cm^2$, generate an AP after switching on the signal. Increased current strengths produce oscillations with increasing frequency; however, the typical shape and size of an AP is reached only within a certain range.*

Computation with HH standard data; the numbers at the traces account for the stimulating current densities [in $\mu A/cm^2$]; simulation time: $40\mu s$.

Fig. 9.3 and Fig. 9.4 illustrate the behavior of a squid axon when stimulated from a monopolar point electrode with different current strengths. The solution will be symmetric to both sides. Fig. 9.3 (C) shows the reaction of segment 0 (below the electrode) and Fig. 9.3 (B) that of segment 3. The result is similar to that of the uniformly polarized case (Fig. 9.1): Firing frequency is slightly increased for stronger signals, but for $I_{el} = -15mA$ the curve bifurcates [Fig. 9.3 (C); case 5]. For higher intensities ($I_{el} = -25mA$; case 6) only one AP propagates, as seen by inspection of segment 3 (Fig. 9.3 (B)). In case 5, we obtain two propagating spikes. Note, that the aperiodic reaction to high stimulus strengths is not a fatigue effect, but actually results from

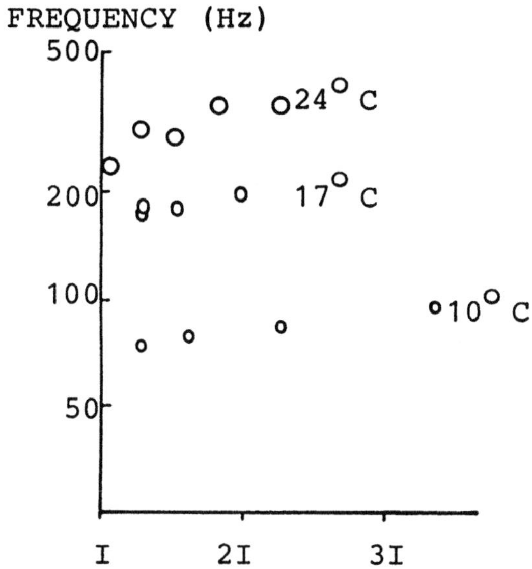

Fig. 9.2 *Frequency-current relation of the uniformly polarized squid axon obtained from experimental data. I is the threshold current which generates repetitive APs. Firing is examined for three temperatures (10° C: small circles; 17° C: medium circles; 24° C: big circles).*

(After Guttman and Barnhill, 1970)

a strong hyperpolarization at the outside segments, according to the negative parts of the activating function.

We can find a similar situation to that of Fig. 9.3 when bringing the electrode closer to the axon ($z = 5mm$; Fig. 9.4), but the range where repetitive firing occurs becomes smaller now.

With the help of equation (7.5) we find that we can also use the results of Fig. 9.3 and Fig. 9.4 for axons with other diameters.

$$\dot{V}_n = \left[-i_{ionic} + \frac{d \cdot \Delta x}{4\rho_i \cdot L} \cdot \left(\frac{V_{n-1} - 2V_n + V_{n+1}}{\Delta x^2} + \frac{V_{e,n-1} - 2V_{e,n} + V_{e,n+1}}{\Delta x^2} \right) \right] / c_m \tag{7.5}$$

In the case of unmyelinated fibers, the active length L of the membrane is equal to Δx. We will obtain exactly the same

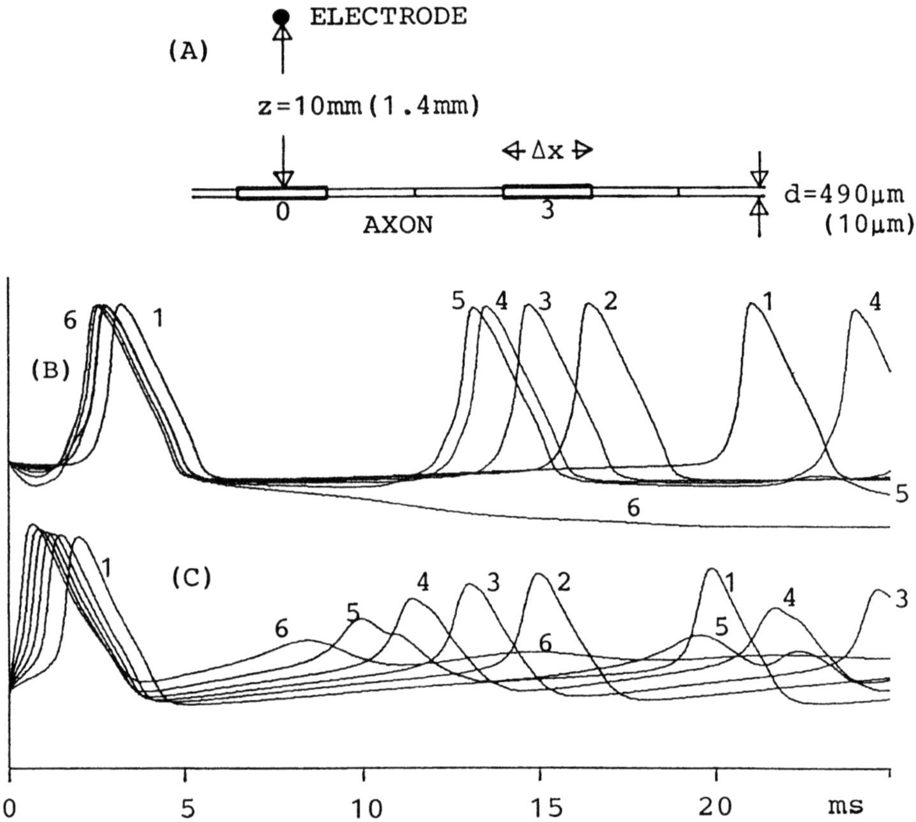

Fig. 9.3 *HH-axon stimulated with constant currents of varied strengths. (A) Geometric situation for two sizes of diameters and distances (smaller measures in brackets; d is divided by 49, all other measures are divided by 7) which numerically produces exactly the same results in the segments when current strength is also divided by 7 (see text).*

(C) and (B) show the answers of segment 0 and segment 3 respectively. The numbers 1,2, ... 6 mark the results for $I_{el} = -3mA$, -5mA, -7mA, -10mA, -15mA and -25mA. For the small sizes these values should be divided by 7. Only the cases 1-4 produce periodic solutions as seen in (B). In case 5, the voltage becomes hyperpolarized after the second peak; in case 6 hyperpolarization occurs even after the first AP.

Simulation with HH standard data, $\rho_e = 300\Omega.cm$, $\Delta x = z/2$; 10 segments.

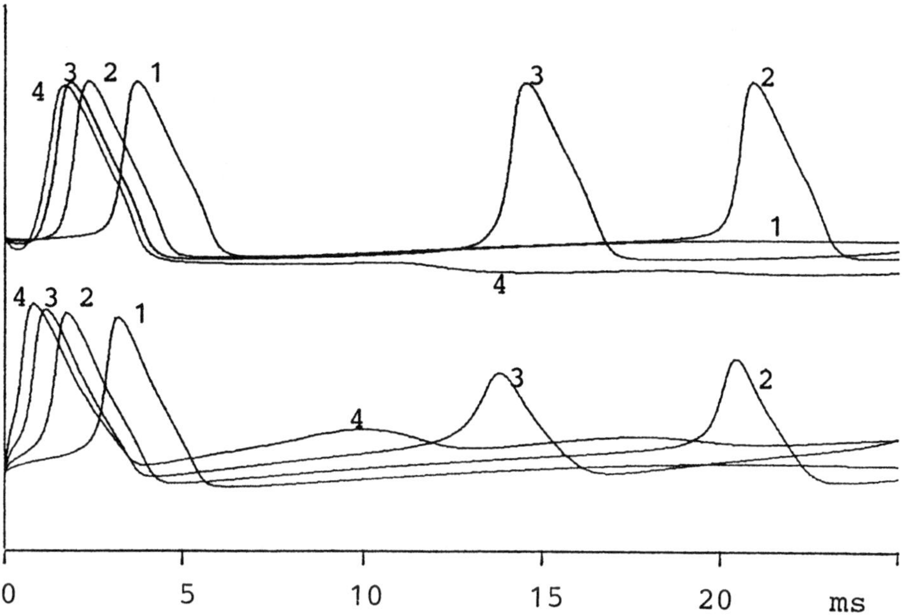

Fig. 9.4 HH-axon stimulated with constant currents. Situation as in Fig. 9.3, but z=5mm. The numbers 1, 2, 3, and 4 mark the results for $I_{el} = -0.5mA$, -1mA, -2mA, and -4mA

numerical results for two different sets of parameters, if $\frac{d}{\Delta x^2}$ has the same value and also the extracellular potential $V_{e,n}$ is the same for corresponding segments. Thus, for a diameter $d_1 = k \cdot d$ we will use $\Delta x = \sqrt{k} \cdot \Delta x$, $z_1 = \sqrt{k} \cdot z$, and $I_{el,1} = \sqrt{k} \cdot I_{el}$, which allows us to obtain the same time-course in the new segments. For example, the curves numbered with 3 in Fig. 9.3(B) and 9.3(C) are calculated for $I_{el} = -7mA$, $z = 10mm$, $\Delta x = 5mm$, $d = 490\mu m$. We will now calculate the parameters needed in order to get a corresponding result for $d_1 = 10\mu m$. Because $d_1 = \sqrt{\frac{1}{49}} \cdot d$, we find $k = \frac{1}{7}$, $z_1 = \frac{10}{7}mm$, $\Delta x_1 = \frac{5}{7}mm$, and $I_{el} = -1mA$. These will produce the same results as above for all the segments. Note, that the propagation velocity also is reduced to $\frac{1}{7}$, which is in agreement with the quadratic relations between velocity and diameter (compare Fig. 6.7).

The repetitive answers of nerve membranes to constant currents are very sensitive to physiological parameters. This is demon-

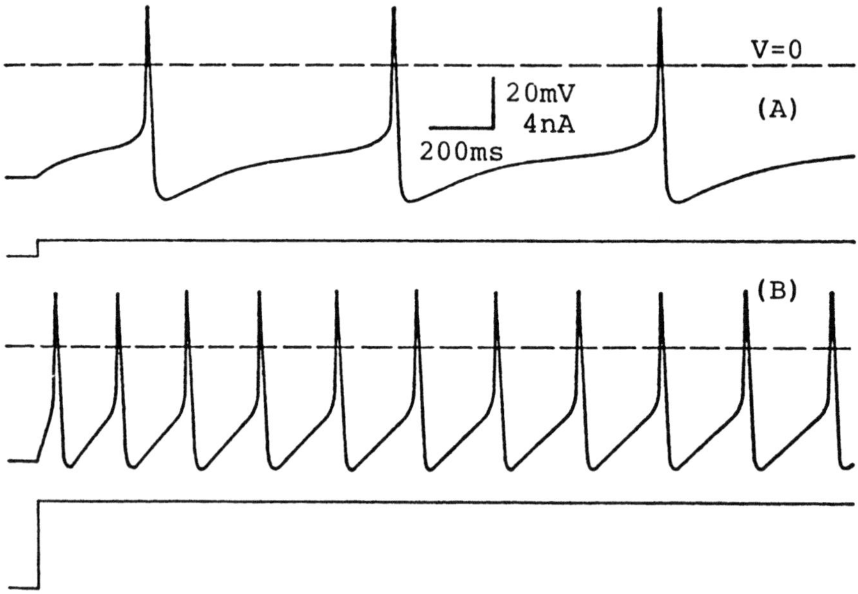

Fig. 9.5 Repetitive firing of a gastropod nerve-cell soma in response to two different constant current stimuli. The upper traces are membrane voltages (dashed line marks zero voltage, not resting voltage) and the lower ones the stimulus signals.

(After Connor and Stevens, 1971a)

strated in several experiments and, if standard data is used, it is not possible to obtain trains of spikes by the FH model or by the CRRSS model.

Stimulated somata will show larger ranges of firing frequencies, than the squid axon membrane, when other channels are involved. CONNOR & STEVENS (1971 a,b,c) made a voltage clamp analysis for *gastropod* nerve cells which respond repetitively to steady current injections (Fig. 9.5). The current-frequency relation is almost linear over a fairly larger range than in the squid axon. The analysis showed that the kinetic properties of a second type of potassium channel are very important in controlling the repetitive firing at low frequencies.

Bursting

Another interesting phenomenon is *bursting*, where periods of firing are separated by periods of silence. Bursting is possibly a result of the neuron activity itself, which changes the inside and the outside ion concentrations and consequently, the membrane behavior. As an example, the calcium concentration can effectively influence firing behavior. If we assume that calcium-dependent potassium channels are involved in the membrane kinetics and that the calcium inflow is coupled with firing, then, after a while, the inside calcium concentration becomes high enough to stop the firing process. Now, the pump- and sequestration mechanisms return the inside calcium concentration to their base levels. This reduces the conductances of the calcium dependent potassium channels and the nerve fiber starts to fire. Such a mechanism has been proposed by GORMAN & THOMAS (1978) to explain bursting in a cell which is often used for experiments, namely the R15 cell of *aplysia* which has been modeled by PLANT (1978). This bursting phenomenon is active without electrostimulation, but it can be influenced by the addition of current.

SCRIVEN (1981) modeled repetitive firing and bursting by using modified HH equations together with the simulation of activity of both the sodium-potassium and the chloride pumps. Besides an additional calcium-dependent potassium conductance, he used a diffusion barrier in the periaxonal space according to FRANKENHAEUSER & HUXLEY (1956): The volume immediatly adjacent to the excitable membrane may be separated from the bulk extracellular fluid by a barrier that limits the ionic movement. With his complex model, SCRIVEN showed that adaption can occur in two ways: Na-K pump activity may increase because of periaxonal potassium accumulation or intra-axonal sodium accumulation; or from increased calcium-dependent potassium conductance caused by calcium accumulating within the axon.

Stimulations with constant currents are interesting from the scientific point of view, but they are not very useful for medical applications. It was mentioned above that extracellular constant current stimulation of nerve fibers would work within a small frequency range for unmyelinated fibers; however, the FH-model for myelinated fibers would fail in the generation of periodic solutions.

Furthermore, charge accumulation leads to gassing at the electrode surface, the positive and the negative electrode currents

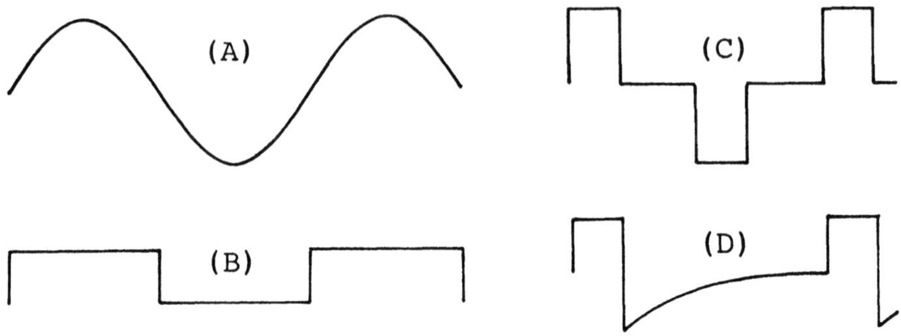

Fig. 9.6 Common signal forms for periodic stimulation. The positive and the negative parts of the signal should be of the same size in order to avoid cell demage when using implanted electrodes. (A) sinusoidal signal, (B) periodic square impulses without gaps. (C) periodic impulses with gaps between the impulses need less charge to reach threshold. (D) exponential compensation of the stimulating pulse is often used because it is technically simple to realize. In order to obtain smaller thresholds, gaps can also be introduced between the square impulses and the onset of the exponential sections.

should be balanced and long unipolar stimulation intervals should be avoided when implanted electrodes are in use.

Periodic stimuli - periodic responses

Fig. 9.6 shows the signal forms used mostly for periodic stimulation. Signals of continuous shape are used especially for cochlear implants, whereas for other applications pulsatile signals are preferred.

Before going into details, we will summarize the most important results which occur by changing the frequency:

i) At very low frequencies, the situation is similar to that observed during the stimulation with constant current. Multiple firing can occur within a half period of the stimulating signal, especially for stronger stimuli.
ii) Within a certain range of frequency, the axon is able to synchronize the AP with the stimulation signal. For intracellular stimulation the excitation synchronizes with the maxima of

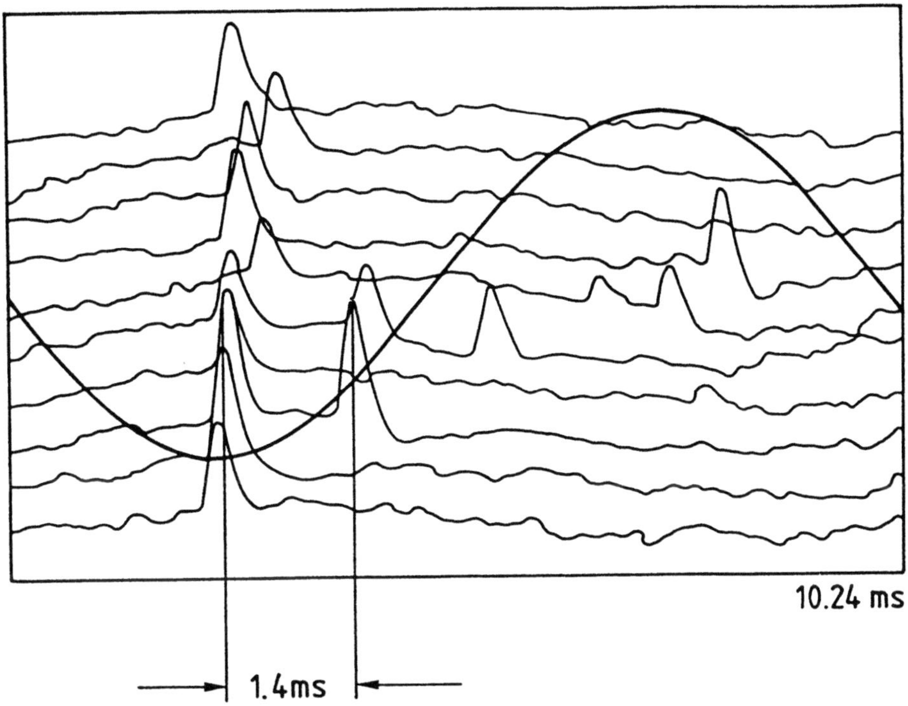

Fig. 9.7 The responses of an axon within the cat cochlear nerve, when it was stimulated with a sinusoidal signal of 100 Hz at the round window. Multiple firing with time intervals of 1.4-1.5 ms can occur. Ten recording sweeps, produced by a 70 μA stimulus signal, are superimposed.

(After Hartmann et al, 1984)

the injected current. In the case of extracellular stimulation synchronizations generally are with the minima of electrode signal. If the stimulus is not strong enough several of those peaks are lost.

iii) Stimulation with higher frequencies needs higher signal amplitudes and synchronization is lost. Switching on the signal causes an AP, but further firing is only seen within a small range.

Fig. 9.7 shows phaselocked AP's which were evoked by sinusoidal electrostimulation of an auditory primary neuron of an intact cochlea. HARTMANN *et al.* (1984) reported that normally one AP was evoked but double and even triple discharges could

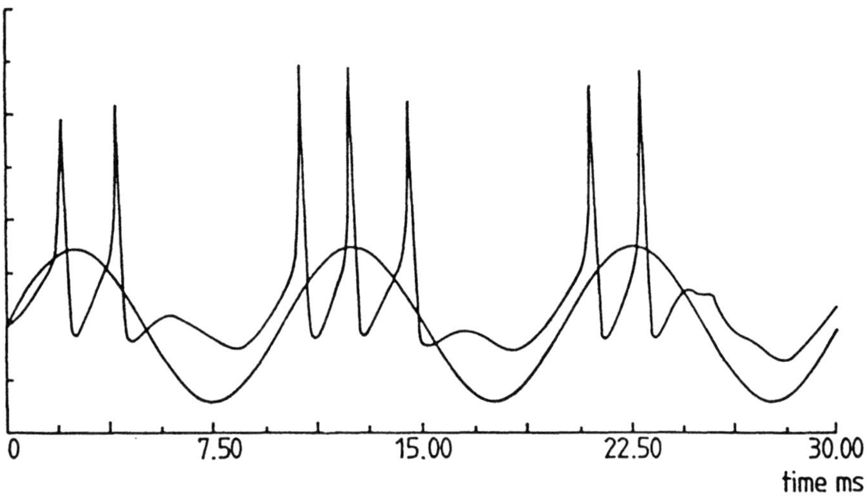

Fig. 9.8 *Stimulation with a 100 Hz sinusoidal signal. Double and triple firing occur during the stimulating half period when simulated with the HH-model. Simulation was done according to the space-clamped model data, but with the temperature factor k=12 in order to get a delay of 1.4 ms between two consecutive AP's. This result was expected from the experimental data shown in Fig. 9.7.*

(After Motz and Rattay, 1986)

occur within one stimulus cycle (as seen in the figure) if the stimulation frequency was low. The nerve responses, shown in Fig. 9.7, are obtained by superposition of ten subsequent answers from one axon, but we would get a similar picture if we compared the activities of ten neighboring neurons in parallel, within a period of the stimulation signal. The response from the whole nerve is responsible for the quality of perception. We will discuss firing patterns in greater detail in Chapter 12.

Fig. 9.8 - Fig. 9.10 illustrate the activity of axons when they are stimulated by sinusoidal signals. Since simulation was done with the space clamped HH-model (Box 4.3), the firing times are synchronized with the maxima of the stimuli.

Even without stimulation, activity exists in the axon, mostly in the subthreshold range, but sometimes the 'spontaneous activity' also produces action potentials. The spontaneous activity was first modeled by HOCHMAIR–DESOYER *et al.* (1984) by adding a noise term in the BVF-model. Addition of noise gives a more

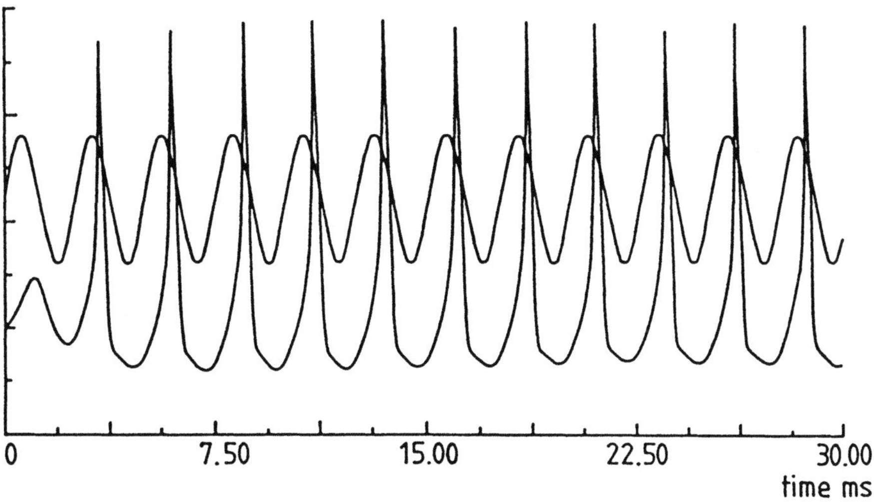

Fig. 9.9 Action potentials produced by a sinusoidal stimulus of 400 Hz. Simulation data as in Fig. 9.8; current strength 20 $\mu A/cm^2$.

(After Motz and Rattay, 1986)

realistic nerve response which is in accordance with single-fiber measurements. Experiments of ROSE *et al.* (1967) and BRUGGE *et al.* (1969) show that the response of electrical stimulated nerve fibers is statistical (see Chapter 12). If a harmonic stimulus is used, the output is more or less phase-locked, but not every amplitude maximum produces a spike. Simulation without using noise will show to less variance in the synchronization and in the sequences of firing. Fig. 9.10 shows the computed reactions of an axon stimulated with a 400 Hz sinusoidal signal.

At first, the simulation of mammalian nerves with the Hodgkin-Huxley equations does not seem to be successful, but it was shown that this model [using a temperature coefficient $k = 10$ or $k = 12$ (see Box 4.3)] is more in agreement with experimental results on the stimulation of warm blooded myelinated nerve fibers than the FH model* (MOTZ & RATTAY, 1986). Fig. 9.11 demonstrates that the FH model fails to work when used to describe the multiple firing reaction at low stimulus frequencies. This is the case

* Since the myelinated parts of the primary acoustic nerve lie close to the electrode, for the intracochlear prothesis, we must assume that excitation is started at these myelinated fibers.

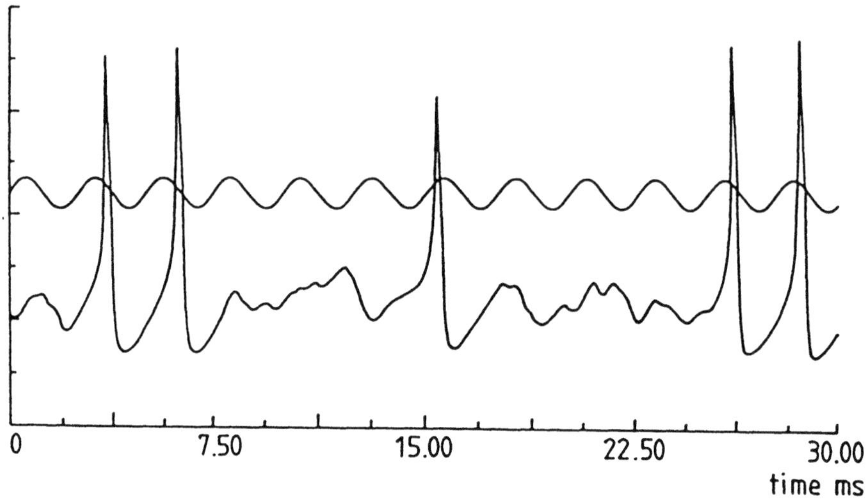

Fig. 9.10 Reaction of the HH-model to a weaker 400 Hz stimulus. A noise signal was added to the stimulus signal in order to simulate the subthreshold activity of the axon which produces sometimes AP's even if no signal is present (spontaneous activity). Because the signal (10 $\mu A/cm^2$) is weaker than in Fig. 9.9, several maxima of the stimulus do not produce spikes.

(After Motz and Rattay, 1986)

even with a high strength.

Apart from most nerve models failing to show this multiple firing process at low frequencies, all the models show similar qualitative behaviors at frequencies higher than 100 Hz. In addtion to the influence of frequency there are also some differences in the firing times when the signal shape is changed (square pulses, triangles, or sinusoidal stimuli), but we will be concerned with these properties in Chapter 12.

Weak superthreshold signals can produce quasiperiodic trains of spikes, but some of the maxima of the stimulus are missed. The numbers of missed spikes are not finely graduated in the case of higher frequencies, as seen in Fig. 9.12(B) for a 2 kHz stimulus. We will see in Chapter 12, that, when a weak signal is applied, the experimental data will show a greater amount of missed spikes than that produced by the BVF-model .

Comparison of the signal strengths in Fig. 9.12 and Fig. 9.13 show that higher frequencies require higher amplitudes to reach

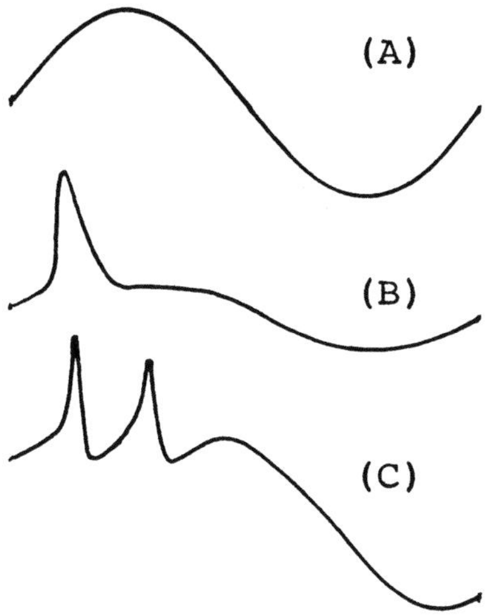

Fig. 9.11 Comparison of the FH and HH-model when stimulated with a sinusoidal signal of low frequency. A 100 Hz sinusoidal stimulus (A) will generate only one AP in the FH simulation even when a high stimulus strength is applied (B). However, double firing will occur if calculation is done with the HH-model (C). The strong negative excursion of (C) is a consequence of the high stimulus.

Calculation was done according to the original FH model and with HH standard data, but with k=12 to obtain (C).

the threshold. This effect also occurs when square pulses are applied. Table 9.1 lists the threshold amplitudes of single square pulses, biphasic, and periodic square pulses without gaps between the pulses. Computations were done with the HH, FH and BVF models. In the case of the HH model, simulation was also carried out for sinusoidal stimulation.

The threshold values for myelinated fibers are calculated with both the original and the corrected FH models. The results are also presented in Fig. 9.14. When long impulses are applied, thresholds are nearly the same for single, biphasic and periodic signals. For short pulse durations the differences in thresholds become more dominant. This situation is common to all models presented in Table 9.1, but it is more striking in the models with slower gating processes. Therefore, the (slow) original FH model needs a higher threshold than the (faster) corrected model. This behavior is documented in Fig. 9.14 (B), where the original model

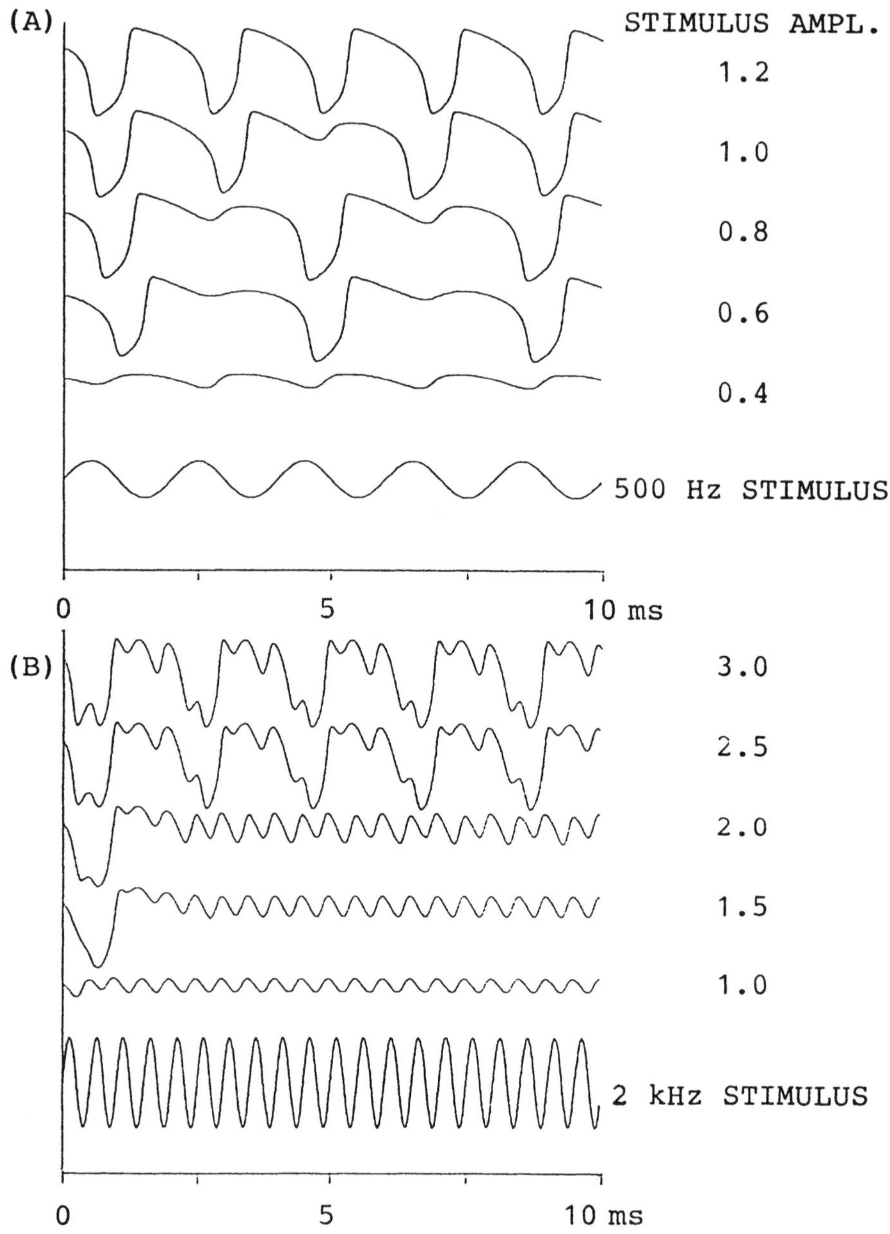

Fig. 9.12 Reactions of the Bonhoeffer-van der Pol-Fitzhugh model to sinusoidal stimuli of different strengths. (A) 500 Hz and (B) 2000 Hz simulated with a local BVF model with data according to Box 5.1, but $\beta = 4$. Note that AP's go downwards.

values are displayed in percent of the corrected one.

High frequency blockade

Pulses with durations in the length of the recovery time, about $1ms$ in mammalian axons, need almost the same current amplitude regardless of being monophasic, biphasic, or periodic. For shorter pulse durations, the second pulse of the biphasic or periodic signal inhibits the formation of an AP; whereas, the third pulse of the periodic signal, having the same sign as the first, again favors AP generation. The difference in the thresholds of monophasic and biphasic signals tends to increase for shorter pulses.

Fig. 9.15 again illustrates the effects of increasing frequency: If stimulation signal periods are shorter than the refractory period, the frequency of firing must be smaller than that of the electrode signal, even for strong stimuli. Further increase will produce a single AP when the signal starts and high stimuli will cause strong oscillations, but we do not recognize the typical shape of an AP. As seen from the phase plane, (Fig. 9.13 (B), Fig. 5.4) the stimulus signal can force the membrane potential to produce relatively large periodic excursions, but the separatrix is not reached because the center of oscillations is lifted in the phase plane (Fig. 9.13 (B)). Enlarging the stimulus produces greater oscillations, but it also creates a higher oscillation center, thus the separatrix cannot be reached again. The amplitude of the oscillations can become greater than an AP when strong stimuli are applied, but even in these cases, normally further spikes will not be produced and the oscillations will become smaller and vanish when we observe the fiber at greater distances from the point of excitation (Fig. 9.16).

There are two situations known in high frequency stimulation which cause repetitive firing: First, even small disturbances in the stimulating signal can produce additional firing as seen in Fig. 9.15 (D). Second, within a certain amplitude range, the system is self-excited, i.e., it produces periodic AP's with a period which, for a given signal frequency, is low at low amplitudes and high at higher amplitudes, ranging from 250 Hz to 1 kHz for the HH model for warm–blooded unmyelianted axons, and from 350 to 600 Hz for

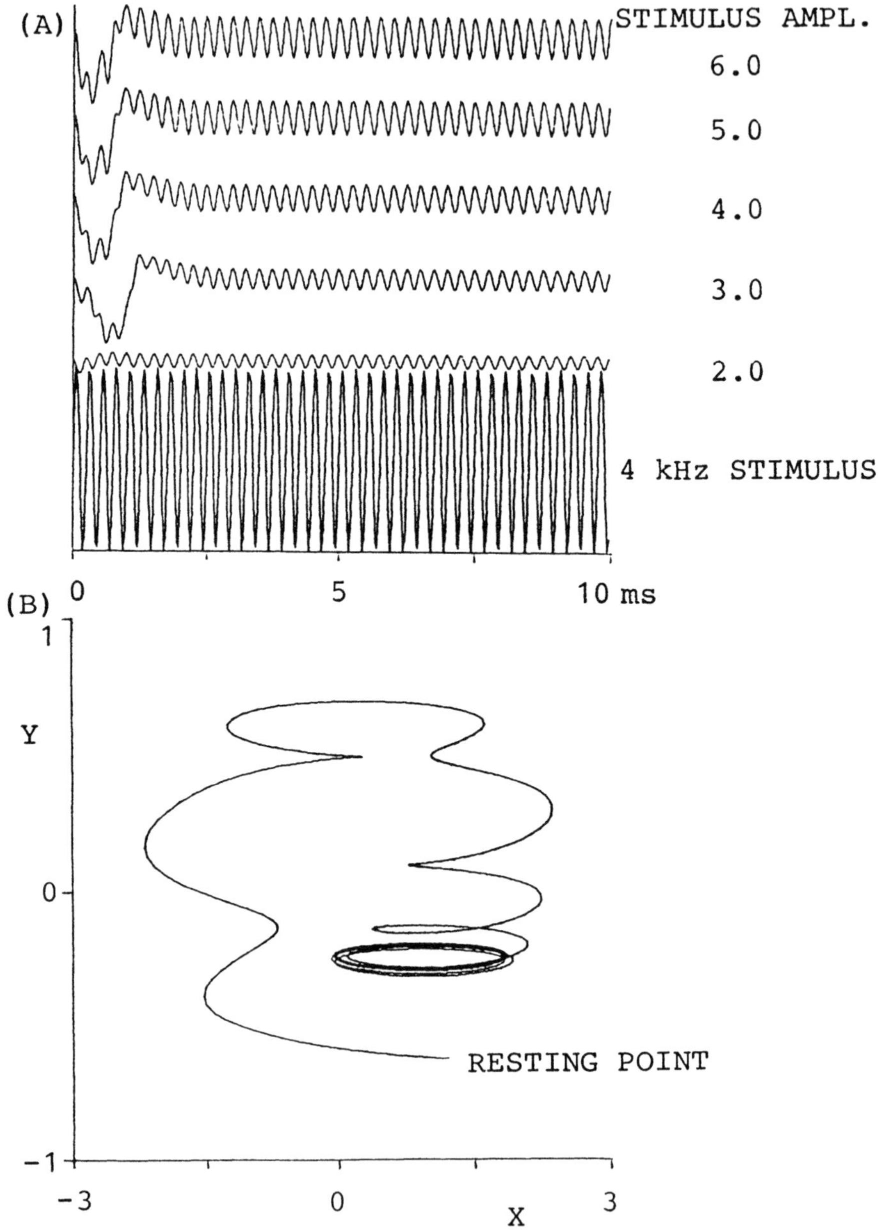

Fig. 9.13 *Reactions of the BVF model when stimulated with a 4 kHz signal. (A) Time courses for different current amplitudes. (B) Phase plot for the case at the top of (A) (with a signal* $s = 6 \cdot sin(2\pi \cdot 4 \cdot t)$*) Data as in Fig. 9.12.*

Table 9.1
Threshold current densities for intracellular stimulation
(space-clamp experiment; currents in $\mu A/cm^2$)

	Time of one pulse in ms						
	1	0.5	0.2	0.1	0.05	0.02	0.01
	corresponding frequency for periodic signals [in kHz]						
	0.5	1	2.5	5	10	25	50
stimulus							
	unmyelinated fiber — HH-model						
single	18	24	50	96	190	490	1020
biphasic	18	30	60	220	620	3000	10000
periodic	18	30	80	200	500	2000	8000
sinusoidal	30	40	120	270	600	2200	9000
	myelinated fiber — original FH-model						
single	360	407	580	870	1450	3300	6400
biphasic	360	414	630	1000	1950	5700	14700
periodic	360	414	620	970	1700	3400	7200
	myelinated fiber — corrected FH-model						
single	380	390	500	730	1200	2700	5200
biphasic	380	390	520	820	1500	4300	10300
periodic	380	390	520	800	1400	3200	6200
	scaled BVF-model						
single	75	79	105	160	290	725	1300
biphasic	75	79	110	220	660	*	*
periodic	75	79	115	220	440	1100	2200

* not excitable

Simulation according to the standard data – HH model, but for k=12; FH model with Temp=37^oC; BVF model with $\beta = 7$.

In the case of the BVF model, the current densities are derived from the normalized variable of the model by multiplication with a scale factor of 440.

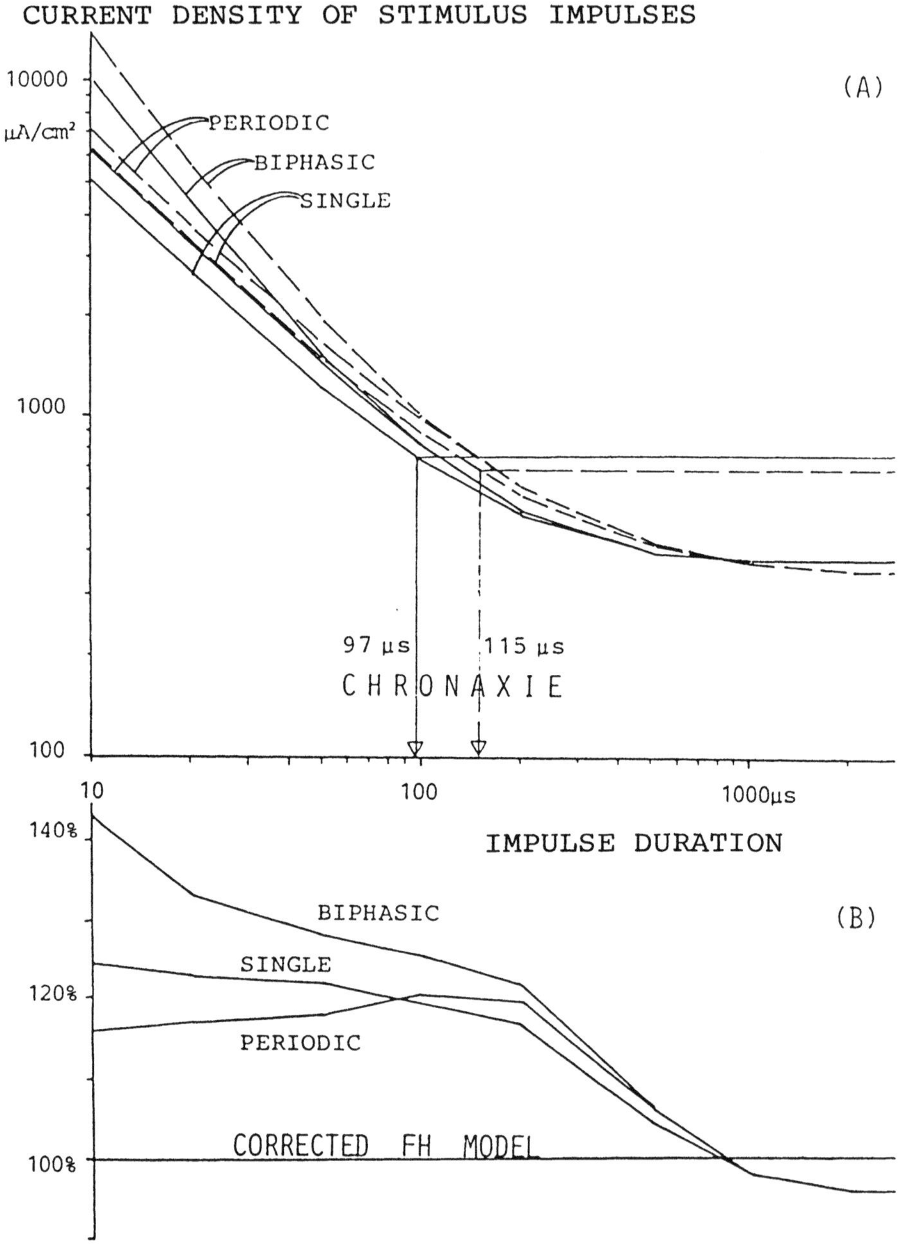

Fig. 9.14 Comparision of the original and the corrected FH models when they are stimulated with single, biphasic, and periodic pulses. (A) Strength–duration curves of the corrected model (full lines) show smaller thresholds for short signals than those calculated with the original model (dashed lines). (B) Relative currents strengths of the original model (corrected model=100%).

the FH model (Fig. 9.17). There is a certain analogy with the well known 'ringing' of axons under the influence of constant current stimulation. The amplitude range within which self-excitated APs are produced, is approximately a constant width, from 2.5 kHz to 12.5 kHz in the case of HH. This becomes narrower towards the upper limit. According to the original FH model, the self-oscillation range extends to much higher frequencies. There is a qualitative but not a quantitative agreement between the two models.

The AP's are not synchronized with the high frequency stimulus phase. There is a continuous change between the synchronization until about 2 kHz (where certain peaks of the signal are missed) and the nonsynchronized AP's at high frequencies. It is evident that it gradually becomes impossible for the self-excitation to follow the amplitude change of the high frequency signal. But, with the help of additional pulses, it is possible to synchronize the onset of the self-excited AP with the stimulating signal. With very short pulses, of $10\mu s$, and with amplitudes of a small percentage of the signal amplitude, which by themselves are subthreshold, synchronization is achieved when the pulse coincides with an upswing of the high frequency signal. In the case of the HH model, the pulse height required to produce an AP beyond 12.5 kHz becomes larger and would lead to an AP if it was applied by itself.

In the next chapter we will discuss a technique which uses another property of high frequency stimulation: an AP, that runs along through an axon, will be blocked when coming to a region which is strongly stimulated with a high frequency signal. This principle is illustrated in Fig. 9.18 and Fig. 9.19, where a single fiber is stimulated at two positions. At the section which corresponds with the lowest lines of the figures, a periodic sequence of 500 Hz AP's is evoked. A certain distance away, there is another monopolar electrode which stimulates the fiber with a sinusoidal signal. The high frequency stimulus generates only one AP when switched on, and this AP propagates in both directions. After about two *ms* that AP, which propagates to the low frequency stimulated part (downwards in Fig. 9.18) collides with the first AP generated there. In contrast to other waves which do not change their shapes after penetration, the AP's extinguish after collision. However, the other AP's, which are evoked by the low frequency stimulation do not reach the upper part (Fig. 9.18);

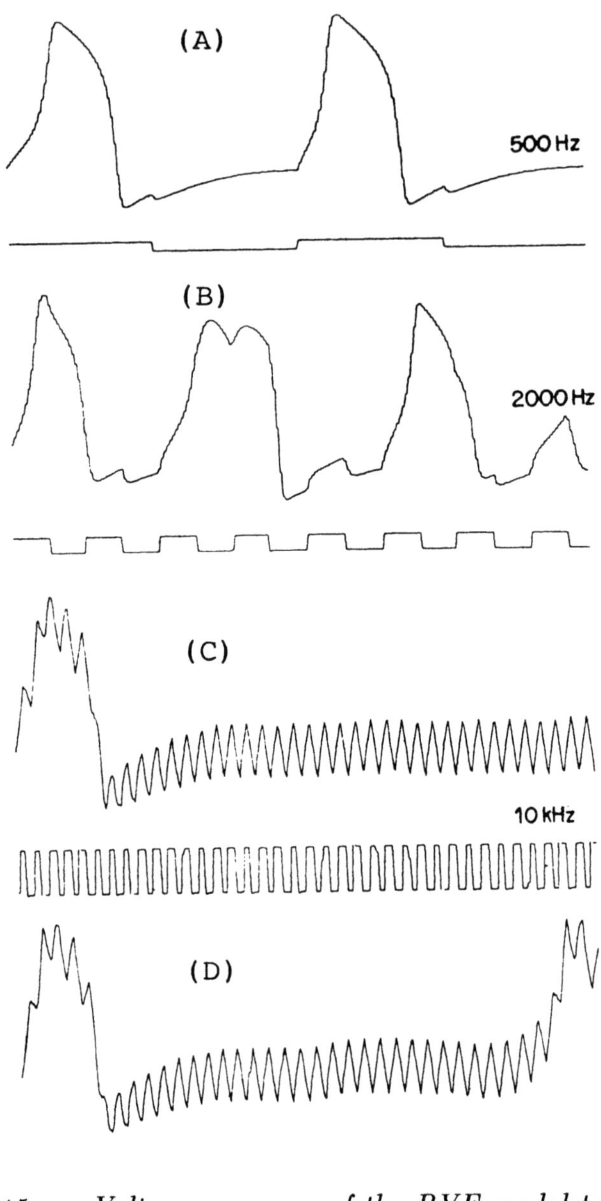

Fig. 9.15 *Voltage response of the BVF model to square pulses with different frequencies. (A) At 500 Hz the axon can synchronize with every pulse maximum. (B) Higher frequencies need stronger currents, but peaks are lost even for high currents. (C) At 10 kHz, an AP is produced only when the signal is switched on. (D) Addition of noise can produce irregular trains of spikes also at high frequencies. Simulation with standard data, but* $\beta = 7$.

(After Rattay, 1986a)

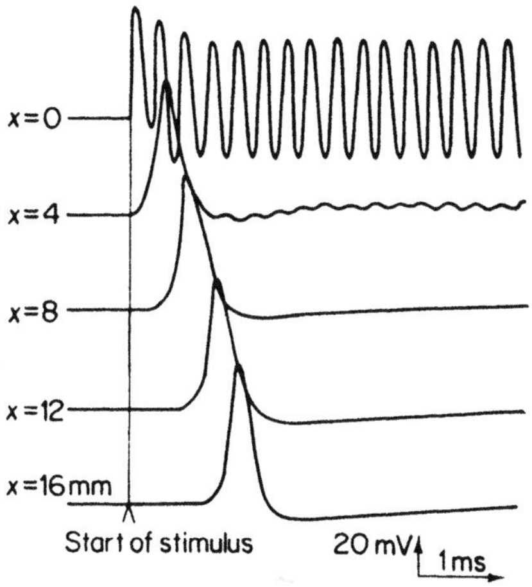

Fig. 9.16 *The fiber response to a 4 kHz square signal injected at the point $x = 0$ where x measures the distance along the axon. The strong oscillations, caused by high stimulus signal, are seen at the point of injection. However, at a distance of 4 mm the potential is oscillating slightly after the first AP, whereas no oscillations are seen for $x \geq 8mm$. Only the switching on effect produces an AP and this AP propagates outward in both directions. Simulation with HH model with $k = 12$.*

(After Rattay, 1986a)

because they are blockaded in the strong oscillating area shown in the center of Fig. 9.18. In this way, only one AP will be sent to the upper end of the axon, namely that which is produced by switching on the high frequency signal, whereas at the other end of the axon a 500 Hz train of spikes will arrive.

If we reduce the amplitude of the high frequency stimulus, some of the low frequency signals will pass. This is illustrated in Fig. 9.19 (A). After further reduction, the effect of the high frequency stimulation is lost; therefore we find no 'switching on' spike and all the low frequency signals will reach the upper end [Fig. 9.19 (B)].

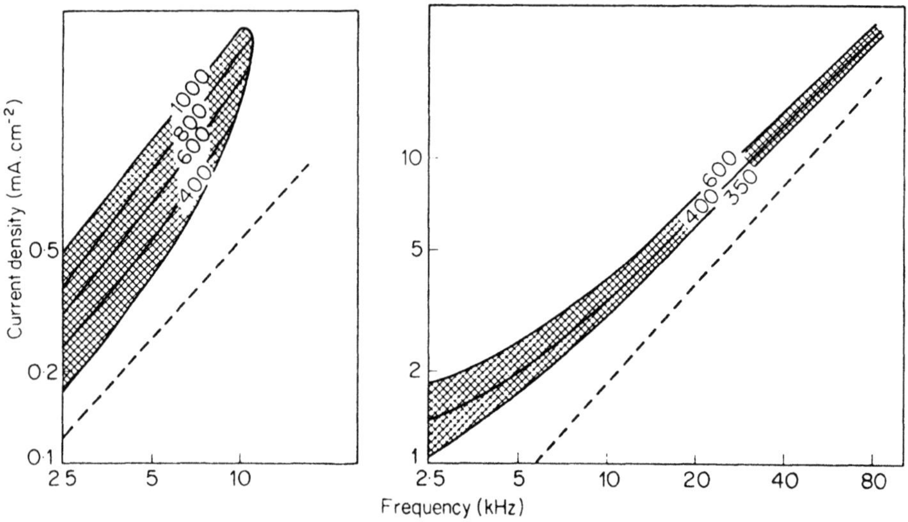

Fig. 9.17 *Firing rates of axons according to the HH model (left) and the original FH model (right) depending on frequency and amplitude of the stimulating signal. Stimulus is a sinusoidal signal of constant amplitude and frequency. The area where firing is possible is shaded. Numbered lines mark the firing rates in Hz. The threshold for switching on the signal is given by the dashed lines.*

(After Rattay, 1986a)

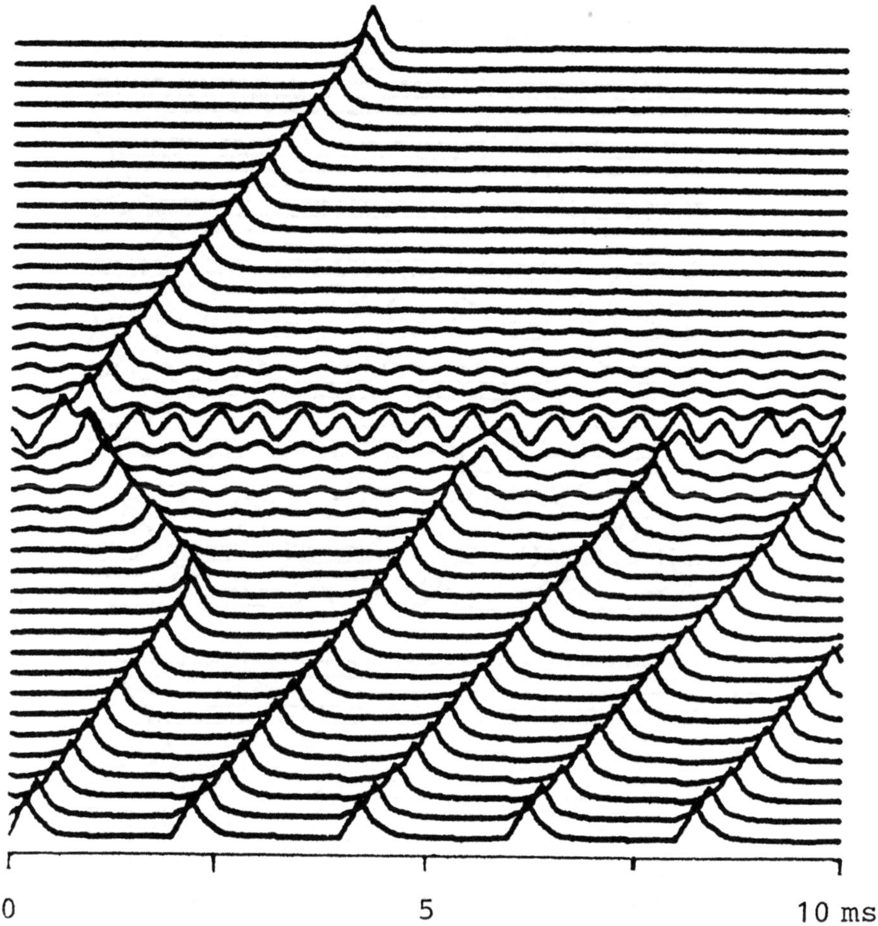

Fig. 9.18 High frequency blocking of spike trains. Simulation of the response of 40 segments of an unmyelinated fiber, which is stimulated at two positions with monopolar extracellular electrodes. At x=0 (lowest trace) square pulses produce a 500 Hz train of APs (that propagates to both sides, but only the upwards propagating part is shown). At x=1cm another electrode stimulates the fiber with a 2 kHz sinusoidal signal. This signal has a maximal effect at the 20th segment. Thereby only one AP is generated and it propagates to both sides. The 500 Hz activity is totally blocked by the high frequency signal but only the 'switching on' effect of the 2 kHz stimulus generates an AP which passes upwards.

Simulation with standard HH data, but k=12, diameter: 10 μm, Δx=0.05 mm, electrode distances 1mm, amplitude of the 2 kHz stimulus: 2 mA.

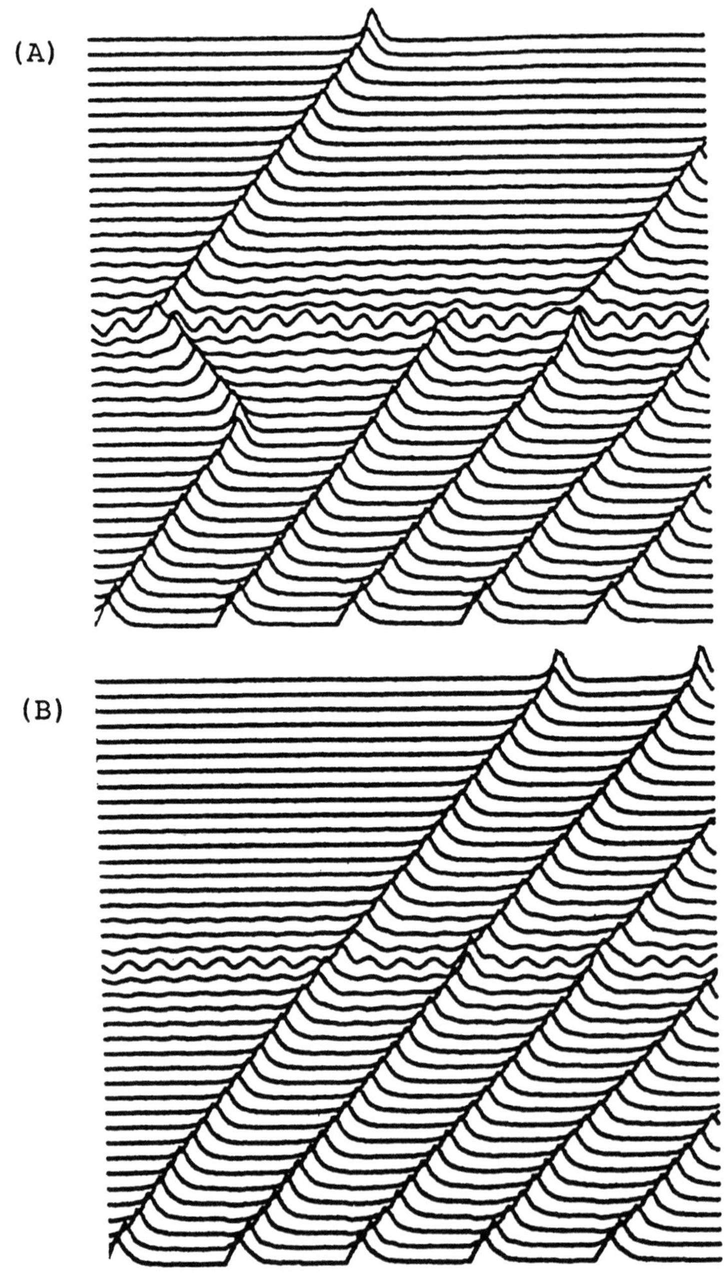

Fig. 9.19 *High frequency stimulation, as in Fig. 9.18, but with different strengths. At a weaker signal (1.2 mA) the blockade partially failed (A). A 1 mA signal of 2 kHz does not influence the firing behavior of the axon.*

10. CONTROL OF THE NEUROMUSCULAR SYSTEM

The inverse recruitment order

In the previous chapters we were concerned mostly with nerve activations in the space domain, but especially for the force control of stimulated muscles we are also interested in the time behavior, because the firing patterns of the stimulating nerve can evoke either smooth or rippled motions.

Fig. 10.1 shows the force response of a skeletal muscle, as a function of stimulus frequency. Firing rates below 25-30 Hz produce contractions which are not smooth enough for most practical purposes. Smooth contractions at high force levels are obtained with higher stimulation frequencies, but if the motor nerve is firing with more than 50 Hz the fatigue effect is very fast (Fig. 10.2; SOLOMONOW, 1984).

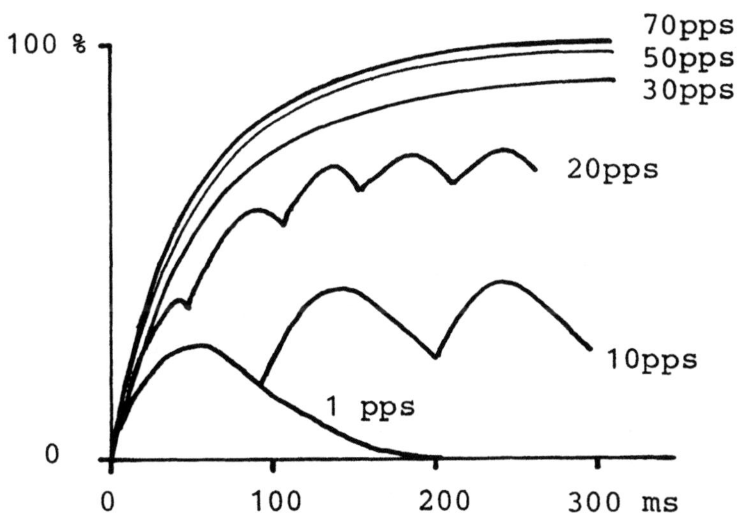

Fig. 10.1 Muscle force as a function of stimulation frequency. The force of a skeletal muscle depends on the pulses per second (pps) which are sent from the stimulating motor neuron. Only some force regulation is available for high stimulation rates. At low stimulation rates the response is not smooth.

(After Solomonow, 1984)

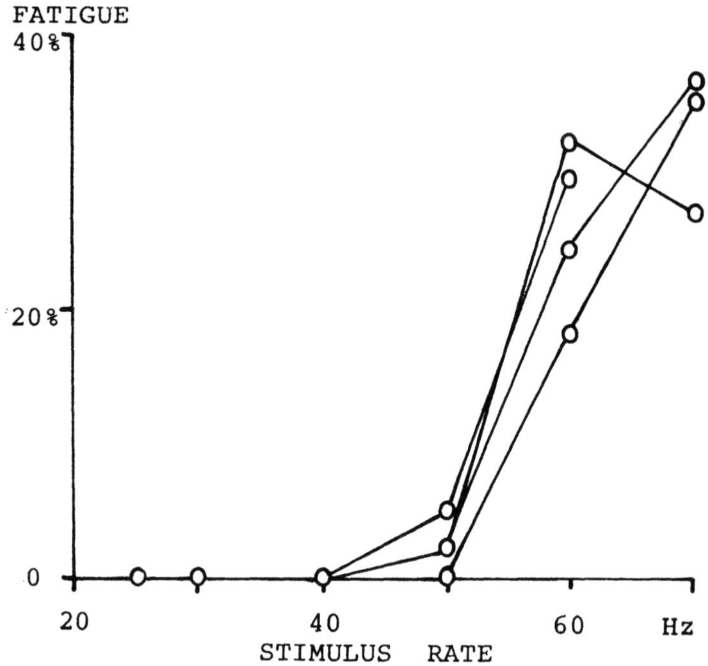

Fig. 10.2 *Muscle fatigue as a function of stimulus frequency. Fatigue is shown as a percentage decrease in force after 30 seconds of contraction. Higher stimulation rates result in a rapid decrease in force.*

(After Solomonow, 1984)

Contractions of electrically stimulated muscles are different from those under physiological control:

i) In contrast to simple artificial stimulation, the motorneurons fire asynchronously during physiologically controlled muscle contractions.

ii) A special motor unit is recruited, after the muscle has contracted to a certain level, when stimulated naturally. These different levels depend on the sizes of the motor units. HENNEMAN'S SIZE PRINCIPLE postulates that larger motoneurons (which control larger motor units) are recruited in the spinal cord when higher contraction levels are reached (HENNEMAN *et al.* 1965a, 1965b; HENNEMAN 1981, KERNELL 1986). The firing rates of the motor units also increase with contraction levels. The efferent neurons innervating the muscles have diameters between 10 μm and 20 μm. According to the size principle the 10 μm axons will be activated at first. But, in the case of electrical nerve stimulation BLAIR & ERLANGER (1933) found the 'inverse recruitment' order: the larger units are activated at first. This is in accordance with the theory (Chapter 8). Further experimental evidence was found, e.g., by PETROVSKY (1978) and FANG & MORTIMER (1987) who measured the widths of single twitch muscle concentrations.

One of the major difficulties in externally controlling the neuromuscular system using electrical stimulation is fatigue. The large motor neurons supply white muscle fibers which, in turn, respond with great force, but they will fatigue fast. In contrast, the small units correspond to red, slow, fatigue-resistant muscle fibers. If a paralized patient should move a glass of water or if he should carry out other light daily activities (e.g., eating, dressing, writing) only the fatigue-resistant small units would be activated.

High frequency blockade

By simple stimulation techniques, pulse durations, pulse amplitudes, or both are varied in order to control the muscle force. But in every case, most of the work will be done by the white muscle fibers, which are activated first, according to the reversed recruitment order. Two techniques are in use in order to obtain the natural recruitment order, where muscle contraction starts with the red, fatigue-resistant fibers, innervated by the smaller motor nerves. Both methods use the same principle: within the nerve, the thick axons are stimulated first, but when high stimulating currents are applied they are also blocked before the smaller fibers. (RANCK 1975; SOLOMONOW 1984; FANG & MORTIMER 1987)

Studies about the high-frequency stimulation of nerve and muscle fibers show this effect, which is kown from Chapter 9 (Fig. 9.17): **the activation of the nerve is blocked by strong high-frequency stimulation and this inhibits the muscle from contraction.** (WEDENSKY 1884; TANNER 1962; WOO & CAMPBELL 1964; MULLER & HUNSPERGER 1967; SOLOMONOW 1984)

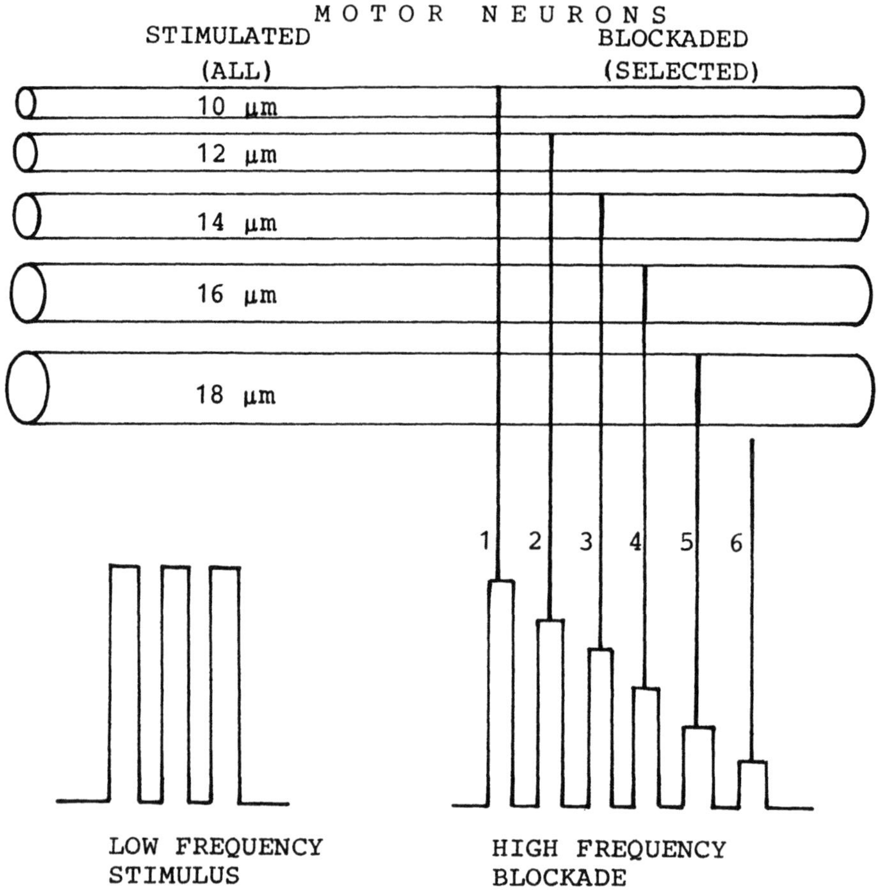

Fig. 10.3 *A natural recruitment order produced by a high frequency block (schematic). Two electrodes were used in order to stimulate the nerve. The lower part of the figure shows the stimulus signals coming from these electrodes. The first electrode will stimulate all the axons in the nerve at frequencies of 25 to 70 Hz. The second electrode is distal and operates with a high frequency but it has modulated amplitudes.*

Strong high frequency signals will blockade all axons (case 1), but when the amplitude is decreased, the spikes can travel across the high frequency stimulated part in the smaller axons. In case 2, all fibers will be blocked, provided their diameters are greater than 12μm. *In case 3, fibers with diameters greater than* 14μm *cannot fire.*

(After Solomonow, 1984)

Since the large motor units are blocked first and since the process is instantaneously reversible without altering the physiological properties of the tissues involved, SOLOMONOW and his group could develop a new stimulation concept. They use two electrodes, one for stimulation, the other for block in order to create a proper spike pattern. The first electrode should produce spikes in all motor neurons of the nerve with a frequency of 20 to 70 Hz. A second electrode, placed distally, changes the potential with a constant frequency but with a varied amplitude (Fig. 10.3). In the figure, the strength of the high-frequency signal is numbered with 1, 2, ... 6. In case 1 (high strength), all the fibers are blockaded. When the strength is reduced (case 2), the signal is not strong enough to block fibers with a diameter less than $10 \mu m$ etc. At a low level, all the spikes in the nerve can pass (case 6). In reality, the behavior of the nerve fibers, as well as, the distribution of fiber diameter is statistical. This will cause some fluctuations in the array of nerve responses, even if periodic stimuli are applied.

We will now assume, that the high frequency amplitude will decrease slowly as shown schematically in Fig. 10.3. Starting with a strong high-frequency stimulus, we will block all the fibers, hence no contraction of the muscle will result. Decreasing the amplitude will allow firing of some axons. As seen in Fig. 9.18 (A), several spikes will bypass the blockade and we can expect that this will occur mostly in the smaller fibers where the influence of high frequency electrodes results in small activating functions, i.e., in fibers which are distant from the electrode. Therefore, the muscle will contract a bit and in the meantime the high-frequency stimulus becomes weaker allowing more of the small motor-neurons to fire. Also, some of the medium sized axons become active, and in the same manner as in natural activation, the large units produce high force as soon as the muscle is contracted over a certain level.

With this method SOLOMONOW (1984) stimulated the *triceps surae* muscle and the *medial gastrocnemius* muscle via the sciatic nerve in cats. For his experiments, high-frequency stimulation with monopolar square impulses of 600 Hz to 900 Hz seemed to be most effective, because within this range of frequencies the maximal reduction of muscle force was seen at a given block stimulus pulse amplitude and width.

High current blockade

In the Chapters 7 and 8, we discussed the influence of the activating function f: at those parts of a fiber where f is positive the stimulation has a depolarizing (activating) effect, and at the parts where f is negative the influence is hyperpolarizing (deactivating). For extracellular stimulation there is always

$$\int_{-\infty}^{\infty} f(x)dx = 0,$$

i.e., for every configuration of electrodes the sum of the activating influences is just the same as the sum of the deactivating influences. This means that: in every fiber there are parts which can be blocked because **spikes cannot propagate through regions where $f < 0$ if the stimulus signal is long and strong enough.**

FANG & MORTIMER (1987) used this principle in order to attain the natural recruitment order. Their first experiments were performed with tripolar cuff electrodes, which were installed on the sciatic nerve in order to stimulate the medial gastrocnemius muscle in the cat. (A discussion on cuff electrodes will follow in the next chapter.) They used quasi-trapezoidal shaped current-regulated pulses with a steep front edge, followed by a plateau of 350 μs and then an exponentially decaying tail.

In Fig. 10.4, twitch force profiles produced by strong, long current pulses are compared with the conventional stimulation scheme. Fig. 10.4 (A) shows the muscle force as a function of stimulus amplitude for long impulses. At current strengths of 2.34 mA or more, nearly all the motor neurons are blocked. Decreasing the amplitude allows some additional fibers to fire in the natural recruitment order; i.e., AP's in small fibers escape first from the blockade. Finally, at a current strength of 1 mA, no blockade is active and twitch force becomes maximal because all fibers are firing. Further decreasing of the current strength reduces muscle force again, but not according to the natural recruitment order; therefore, currents lower than 1mA would not be used in this example by the FANG-MORTIMER METHOD. Fig. 10.4 (B) demonstrates that much smaller charges are needed for the conventional stimulation scheme with short square pulses, but attention must

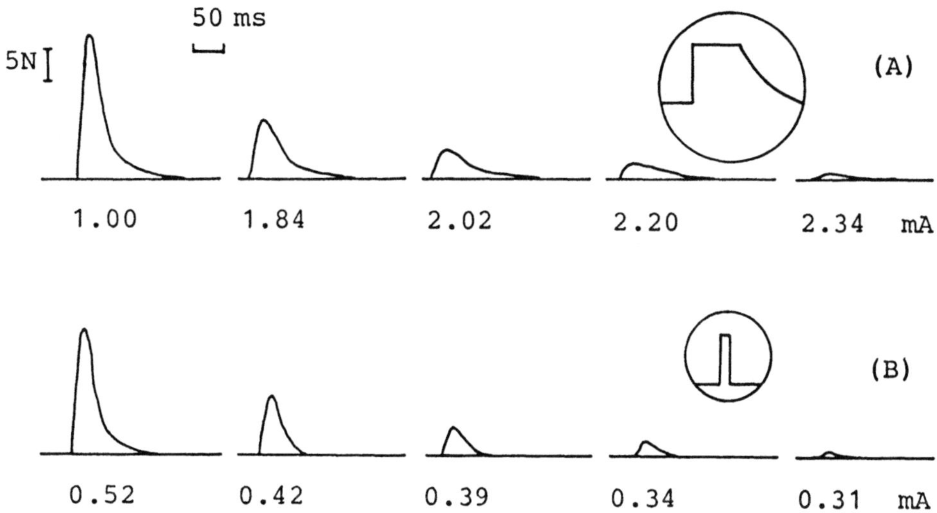

Fig. 10.4 Twitch force profiles for strong and conventional stimulation. Two types of pulses are used for stimulation, as shown in the inserts: (A) with quasi-trapezoidal forms of high intensity, the natural recruitment order is obtained, but muscle force is related inversely to current strength; therefore, the highest stimuli produce the lowest forces. (B) for the conventional stimulation with short square impulses, the relation between the stimulus strength and the force is almost linear within a certain range. However, the recruitment order is reversed; i.e., the large motor units are excited first.

The numbers below the curves show current intensities in mA. Duration of the pulses: 350μs (plateau) in (A) and 10μs in (B).

(After Fang and Mortimer, 1987)

be paid to the disadvantages of the unnatural, inverse recruitment order.

During the fatigue test FANG & MORTIMER (1987) applied 20 Hz stimuli continuously for 2 minutes. The blocking method according to Fig. 10.4 (A) showed a force decay of only 13 % at the end of 2 minutes contraction. In the case of conventional stimulation, this decay reached 70 %.

The round about stimulation

THOMA and his group apply the concept of the 'round about stimulation', which is another method to overcome muscle fatigue. Four electrodes are sewn onto the epineurum of a nerve. By cyclic variation of the active electrodes, different groups of fibers within the nerve are stimulated. The situation is similar to that described in Fig. 8.10, where a dipole stimulates a nerve: even for the highest electrode currents used, only about a quarter of all motor neurons are stimulated. After a stimulation burst containing several pulses, which can be varied in amplitude (e.g., as seen in Fig. 10.5), another pair of electrodes becomes active, thereby other muscle units also will be excited.

If the active (positive-negative) pair of electrodes is perpendicular to the nerve, the central region cannot be stimulated. In practice, the electrodes are sewn in a displaced arrangement anyway; therefore, (in the case of strong stimulation) all fibers of the trunk can be reached within one period of rotation.

The cyclic variation of activated electrodes allows for the stimulation of the phrenic nerve 24 hours per day; therfore, patients can be moved from the iron lung to the wheel chair. The same principle can also be applied to stimulate, for example, the legs of paralized patients. In such cases, two pairs of eight-channel implants are used (Fig. 10.5). By the use of semicustom integrated circuit technology, implanted muscle stimulators have been developed (e.g., SMITH *et al.* 1987; MEADOW *et al.* 1987). The units are small and lightweight, which when combined with low power consumption allows for permanent usage. Most systems operate on similar principles.

The patient command control signal is derived from an appropriate source which is under the voluntary control of the patient, e.g., shoulder - position - movement - command, eye movement control or voice input (SMITH *et al.* 1987; NATHAN 1986). By microprocessor techniques, the information on muscle selection, as well as, for stimulus parameters is derived from the patients commands. This information is sent via a modulated radio frequency to the implanted receiving coil. Most carriers operate at frequencies around 10 MHz. The signal is used not only for the transport of information, but also often for the power supply, thereby becoming independent from the lifetime of implanted batteries (Fig. 10.5). Since the efficiency of such power transfer methods is only

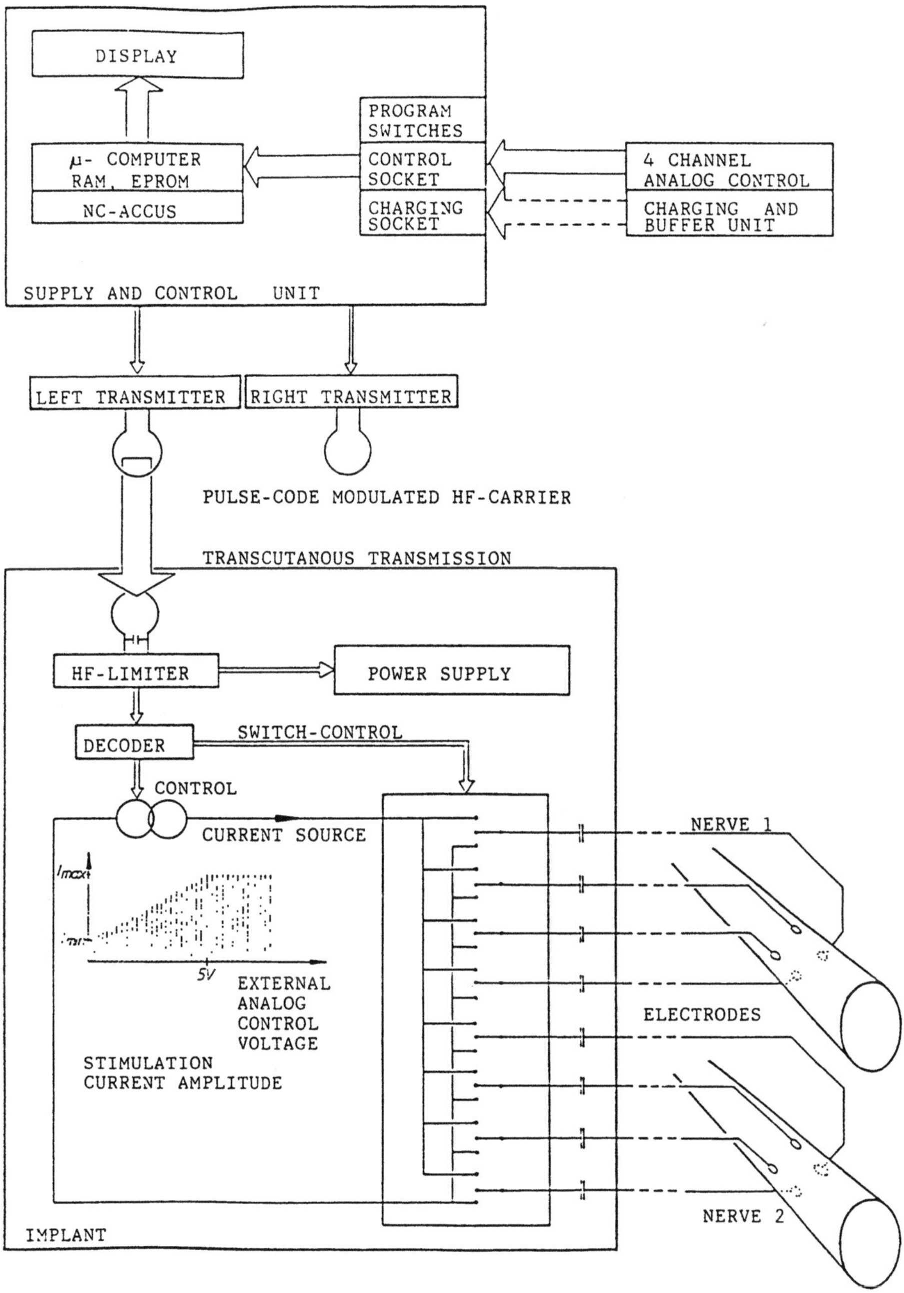

Fig. 10.5 The principle of the 'round about stimulation'. The method is in use for phrenic nerve stimulation as well as for patients with spastic paralyzis.

25-30 %, the circuit power consumption of the implants must be held at a low level (KO *et al.* 1977).

Multi-channel stimulation

Multi-channel strategies allow selective activation of different muscles in order to move, e.g., the upper limbs, to grasp with the fingers or to stand up. The control signal may be used in an open-loop system, i.e., the command signal is directed only under visual control. In closed-loop systems, where feedback is used in the control strategies, more efficiency will be expected. Sensors are used to determine joint angles and forces (CRAGO *et al.* 1986; RUTHERFORD *et al.* 1987) and their outputs are involved in the micro-processor program in order to calculate the next stimulation strengths.

Implanted electrodes are used mostly for muscle activation in paraplegic patients, because of good neural selectivity and the ability to achieve high muscle forces. On the other hand, surface electrodes are advantageous because surgery is not necessary. However, it is very difficult or even impossible to reach the motor neurons and muscles hidden below a thick sheet of fat with surface electrodes. Furthermore, there are often problems in finding the right position and then holding this position when muscles are activated. Special garments have been developed which allow quick fixing of the electrodes without substantial moving during the time of the operation. Besides the reactivation of partially paralyzed muscles and the restoration of gait, multichannel stimulation with surface electrodes has been applied with success for standing and walking, as well as, for hand movements (KRALJ *et al.* 1983; KRALJ *et al.* 1986; NATHAN 1986; VOSSIUS *et al.* 1987).

Electric stimulation is also employed for cases of incontinence, and vaginal or anal stimulation. Vaginal stimulation is used clinically to improve urinary incontinence by enhancing the activity of a weakened urethral mechanism or by inhibiting of overactive detrusor. This may be done by excitation of the pudendal nerve fibers which can be reached with vaginal or anal electrodes.

11. CASE STUDIES: NERVE CUFF ELECTRODES, STIMULATION BY MAGNETIC FIELDS

Split–cylinder and spiral nerve cuff

Cuff electrodes are successfully used in neuroprosthetic applications to activate the lower extremities, the bladder, and the diaphragm. It is also used for the treatment of chronic pain. These electrodes can be positioned so that the relative motion, caused by muscle contraction and limb movement, is minimal.

Fig. 11.1 shows schematically four types of nerve cuffs suited for medical applications. One or more dot or ring electrodes are mounted at the inside of the cuff in order to awake selective neural reactions or unidirectional firing.

Good results are available with minimal diameter cuffs, as long as the intraneural blood flow is not affected. Snugly fitted cuffs will hold their positions and are also suited for dot electrodes. According to RYDEVIK *et al.* (1981), this requires for an internal cuff pressure below 20 $mm\ Hg$ (0.27 N/cm^2). NAPLES *et al.* (1988) has shown that spiral nerve cuffs fit nerves very snugly. A spiral cuff, which naturally has a diameter slightly smaller than the nerve, will automatically conform itself to the nerve. As an example, if a two–wrap spiral cuff made from elastic material of 0.3 mm thickness (as discribed by NAPLES *et al.*, 1988) with 1 mm diameter is placed on a nerve with a diameter of 1.21 mm it will induce a 20 $mm\ Hg$ increase in internal pressure. In contrast, if a 1 mm split–cylinder cuff, like that shown in Fig. 11.1 (B), is placed on a 1.003 mm nerve, the critical 20 $mm\ Hg$ increase in internal pressure is reached.

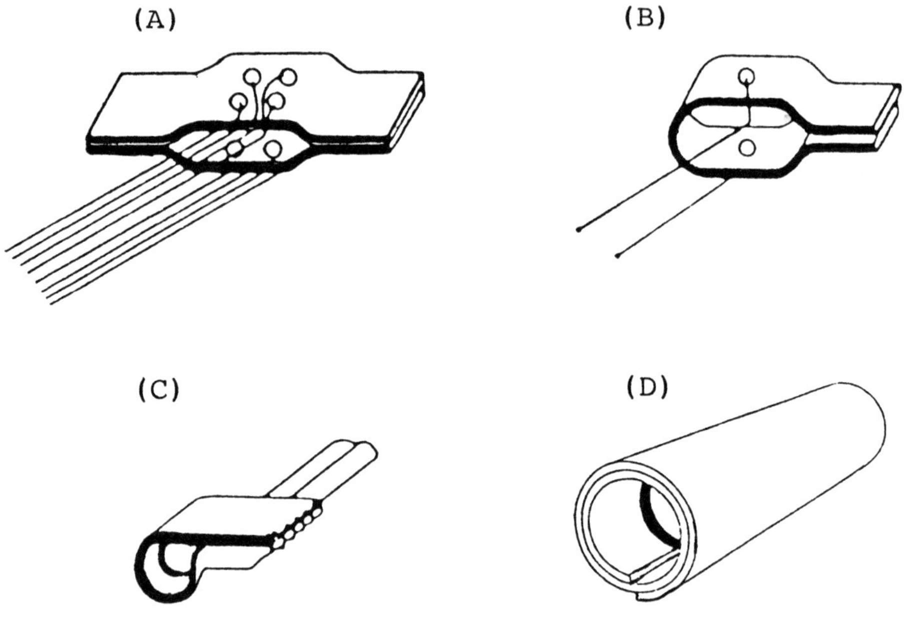

Fig. 11.1 *Nerve cuff electrodes of Avery and Wepsic (A), Avery (B), Hagfors (C) and Mortimer group (D). Blocking cuffs with different diameters like those described by Sweeney and Mortimer (Fig. 10.2) can be created easily by appropriately trimming the spiral cuff (D).*

(After Naples et al., 1988)

One way firing

An interesting application using cuff electrodes, is the generation of unidirectionally propagating AP's. Unwanted efferent motor nerve activity can be stopped by collision block, i.e., a special electrode combination can be used to generate a train of spikes only in the afferent direction and these antidromic AP's annihilate the orthodromic AP's by collision. [VAN DEN HONERT & MORTIMER, 1979, 1981; SWEENEY & MORTIMER, 1986; UNGAR *et al.*, 1986; MORTIMER *et al.*, 1988]

The simplest situation is stimulation using a monopolar ring electrode mounted asymmetrically in a relatively long cuff with a constant diameter. UNGAR *et al.* (1986) investigated this problem and they have found that, even with this simple configuration, it is possible to find a block window, i.e., within an interval of current strength, all the fibers in the nerve trunk will fire unidirectionally. In their research, they used a cuff containing 7 ring electrodes with

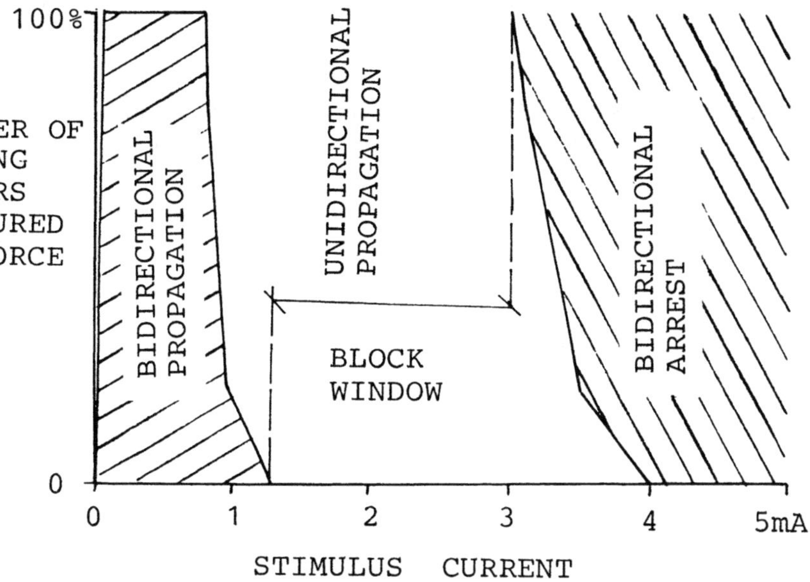

Fig. 11.2 Different firing behavior of motoneurons as function of stimulus strength of a monopolar cuff electrode. A 24 mm cuff with 1.65 mm internal diameter was put on the isolated branch of a cat sciatic nerve (1 mm diameter) which innervates the medial gastrocnemius. A ring electrode was positioned 3 mm from the distal end of the cuff. The evoked muscle force is a measure of the population of firing axons. [In order to find the upper border of the unidirectional propagation a ring electrode 3 mm from the other end was activated.] If the pulse amplitude is within the block window (1.25 – 3 mA) all the axons of the nerve trunk will fire synchroneously in one way.

Stimulation was done with quasitrapezoidal cathodic pulses with a plateau phase of 320 μs and 600 μs falling time. This pulse form needed only a minimal charge.

(After Ungar et al., 1986)

a distance of 3 *mm* between the neighboring electrodes. Beside a mass electrode placed in the back of the animal only one of the electrodes was active.

Fig. 11.2 shows typical experimental results for varied stimulus strengths with an asymmetric electrode position: Weak superthreshold cathodic stimuli generate bidirectional propagating AP's. Higher currents cause the spike to propagate only to the far end of the cuff, and further increases lead to a total arrest (no firing) in all the fibers of the nerve trunk. One way firing in all fibers was achieved in 6 of 8 animals. The best results were found with a high asymmetry (the cathode should be placed at least 5 *mm* from the cuff end) with long quasi-trapezoidal stimuli. The relatively long falling phase is important for unidirectional firing. Strong square pulses and pulses with short falling phases generate an AP by the switch off phase.

Better results are available with assymmetric cuffs as shown in Fig. 11.3. These can be designed with a shorter length, which is more suited for clinical application. (SWEENEY & MORTIMER, 1986)

Modeling of nerve cuff electric fields

The electric fields produced by nerve cuffs can be calculated by iterative, finite difference or finite element methods. Fig. 11.3 shows an asymmetric two electrode cuff designed by the MORTIMER group for one way firing. Note, that the outer electrode has a greater diameter.

By inspection of the isopotentials and the activating function it is obvious that there are only small sections with strong irregularities. They are below the electrodes at the edges of the cuff. Within inner sections (between the electrodes and also between the electrodes and edges) providing that the cuff diameter does not change, the current flows pure longitiudinally. This causes equidistant isopotential lines and zero values in the activation function and in the second difference of the potential (Fig. 11.3).

If the ring electrodes are not to short*, and if the cuff is long enough and snug to the nerve, all the activating influences for all the fibers in the trunk will show a typical form, illustrated by the lower traces of Fig. 11.3. In shorter cuffs or cuffs where the cathode is positioned close to the left end, the hyperpolarizing effect of the left edge would be stronger which causes a smaller block window. We also find a similar situation for big diameter cuffs.

---

* When short ring electrodes or dot electrodes are very close to the nerve trunk they will cause a different reaction in the fibers with a longitudinal set off of their nodes.

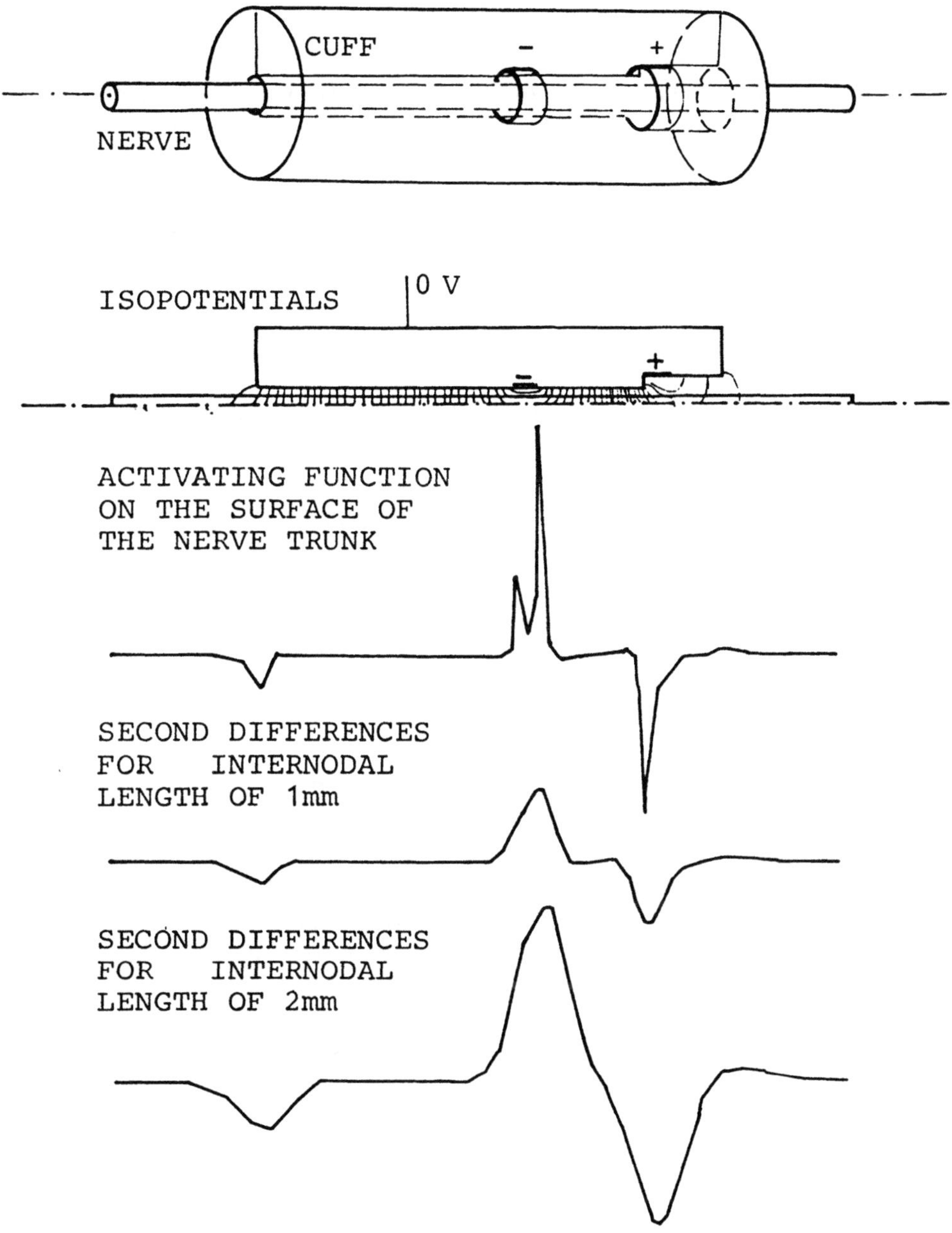

Fig. 11.3 Asymmetric cuff electrode designed for unidirectional firing with second derivative and second difference profiles at the outer edge of the nerve trunk.

(After Ferguson et al., 1987)

A nerve cuff designed for selective excitation

It is highly desirable to have a stimulation system that controls the activation of several muscles with only a single implanted electrode device. In the proximal part of a nerve trunk the axons are highly intermixed respective to their end organ innervation. This is generally not the case near the distal end (before branching of the trunk). In such cases electrical stimulation of one fascicle will result in the selective stimulation of a single muscle (SUNDERLAND, 1978).

In 1990 Sweeney *et al.* have demonstrated by numerical modeling as well as by animal testing that snug nerve cuffs with tripolar dot electrodes are suited for selective excitation. Selectivity is more successful than with monopolar dot electrodes and of course better than that with ring electrodes (Chapter 8). The result is further improved by use of an additional transverse current (Fig. 11.4).

Nerve stimulation with magnetic fields

Since 1985, when BARKER *et al.* introduced magnetic coil stimulation of human motor cortex, this stimulation technique of nerves has been the subject of increasing interest. Magnetic fields are generated by discharging a capacitor into an inductive coil placed over the region of stimulation. Stimulation with magnetic fields has two advantages. First, it is noninvasive and does not require physical or electrical contact between the body and the stimulating coil; second, body structures are more easily penetrated than in the case of transcutaneous stimulation.

The principle of magnetic stimulation is: a changing magnetic field **B** will generate an electric field **E** which can be calculated by the Faradays law (TAY *et al.*, 1989; DURAND *et al.*, 1989; REILLY, 1989; ROTH & BASSER, 1990):

$$\oint \mathbf{E} = \frac{d}{dt} \int\int \mathbf{B} ds$$

Here, the first integral is taken over a closed path and ds is the area element normal to the direction of **B**.

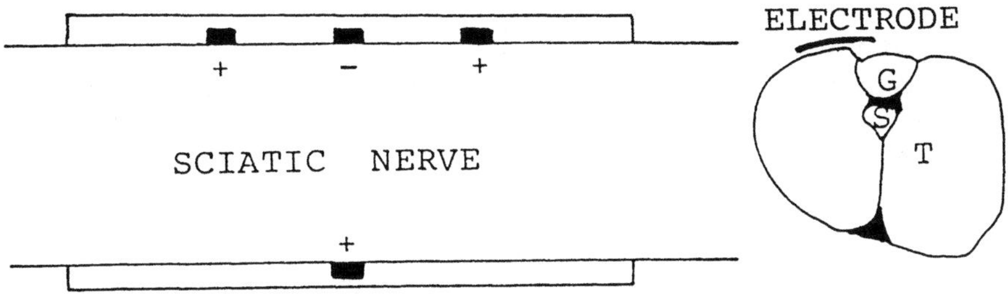

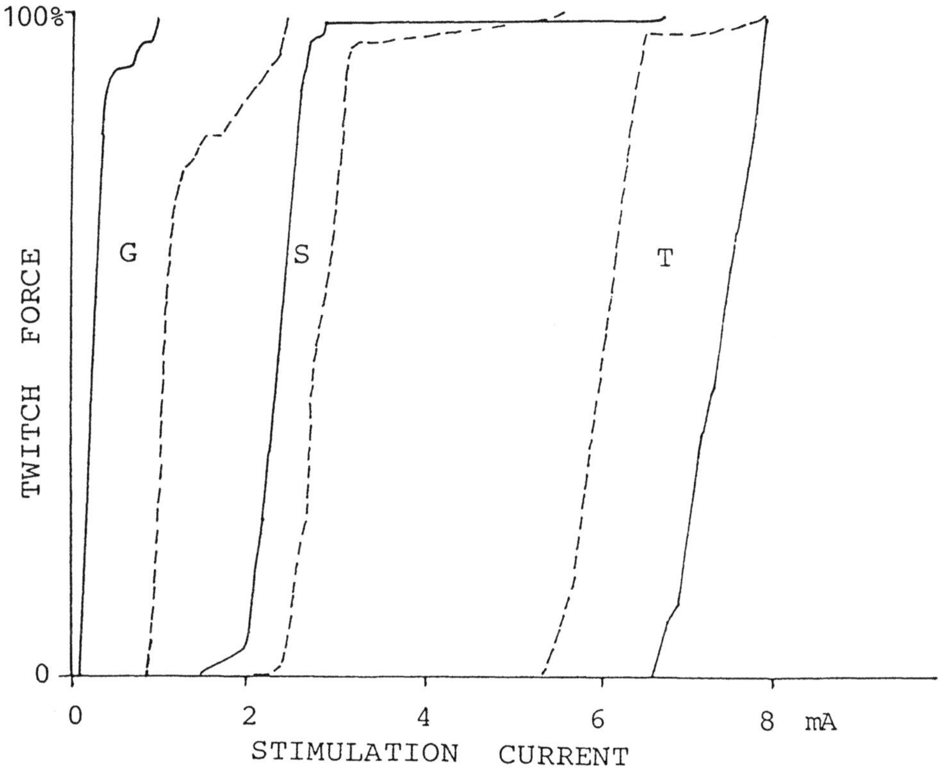

Fig. 11.4 *Selective electical stimulation of a sciatic nerve from a cat. A tripolar cuff electrode (electrode area 1 mm^2, electrode distances 2 mm edge to edge) with an additional anode just opposite to the cathode was used for stimulation of selective stimulation of the medial gastrocnemius (G), the soleus (S) and the tibitalis anterior (T) muscles. Separation by current strengths for tripolar stimulation (full lines) were improved by using an additional subthreshold current of 1 mA from the opposite electrode (broken lines).*

(After Sweeney et al., 1990)

For special configurations of magnetic coils and toroids we can calculate E or we find at least an approximative approach. The coils can be optimized in order to find maximum activating functions, for fibers at defined positions. Besides the spatial influence of the generated electric field, high attention have to be paid on its temporal change.

In practice, usually a capacitor is discharged through the coil, until the current flow returns to zero. Thereby, an electric field is generated, which has a charge balanced biphasic waveform. In 1989 FERGUSON *et al.* showed, that the charge injection at a particular pulse width should be largest in an optimal coil. Thus, by simulation optimal geometric relations can be found.

12. ELECTROSTIMULATION OF THE AUDITORY NERVE - COCHLEAR IMPLANTS

The pioneers

The earliest attempts of hearing by electrical stimulation have been credited to VOLTA in 1790. He connected his own ears with his newly discovered voltage cell and thereby he obtained auditory sensations. Electrical stimulation as a treatment for deafness began seriously in 1957. DJOURNO & EYRIES (1957) reported about acoustic perceptions of a patient when his auditory nerve, which was uncovered by disease and surgery, was stimulated with electrodes. This led to separate efforts of three investigators in California who developed implantable cochlear prostheses, i.e., electrodes, which are implanted into the inner ear. These pioneers (SIMMONS (1966), MICHELSON (1971) and HOUSE (1976)) stimulated other groups in the States, Europe and Australia to develop high quality implants using different coding strategies with the aim to obtain good speech understanding. (PARKINS & ANDERSON, 1983; SCHINDLER & MERZENICH, 1984).

Cochlear prostheses are used for patients who have a mechanical defect in the inner ear. Although their nervous system may be completely intact, if the hair-cells cannot be stimulated acoustic sensations cannot be obtained. In such cases electric stimulation can be used to overcome the defect of mechanical transmission.

In the healthy ear the acoustic signal, which is physically a swinging of air pressure, produces an analogous movement of the hairs of the hair-cells. These motions are finally transformed into neural signal.

Frequency and loudness are the two essential items of the acoustical signal that must be transmitted via the nervous system from the inner ear to the brain. We know that the frequency sensitivity is a function of the place of stimulation, i.e., the hair-cells at the beginning of the cochlea react sensitively to high frequencies whereas only deep tones will reach the hair-cells at the apical end of the hearing organ. Higher sound intensities generate higher firing rates in the fibers of the acoustic nerve, but unfortunately, the general coding system is not competely understood, today.

Although there exists a lot of interesting books and reviews on the natural hearing principles, a short introduction in ear mechanisms and hearing theories is discussed below. [HELMHOLTZ, 1885; RUTHERFORD, 1886; WEVER, 1954; BÈKÈSY, 1960; DALLOS, 1973; MØLLER, 1974, 1983; SCHROEDER, 1975; GELFAND, 1981; STEEL, 1983; ALLEN, 1985; ALTSCHUSTER *et al.*, 1986; LIM, 1986; ABBAS, 1988]

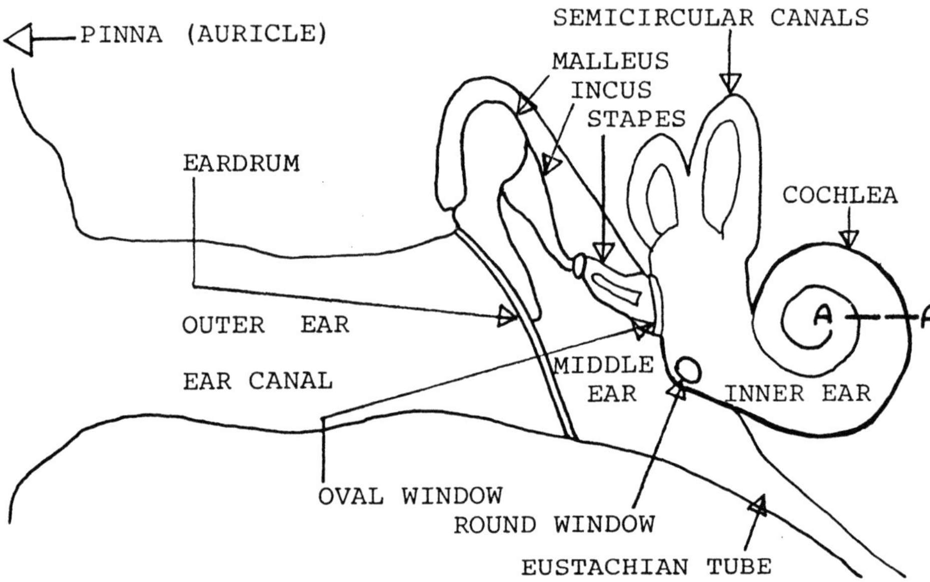

Fig. 12.1 The main parts of the human ear. A cross section (A-A) of the cochlea is shown in Fig. 12.3.

Ear mechanics

The main parts of the human ear are the outer ear, the middle ear and the inner ear or *chochlea.*

The outer ear consists of the *pinna* and the *ear canal (external meatus).* The ear canal is an irregular rather than a simple tube with a cross section of about 7x9 mm and a length of 23 mm. The ear canal ends at the *eardrum (tympanic membrane).* Such a tube resonates at a frequency of 3800 Hz corresponding to a wavelength of 92 mm, which is four times the length of the tube. Damping by the walls and the eardrum causes a boost in the resonance to

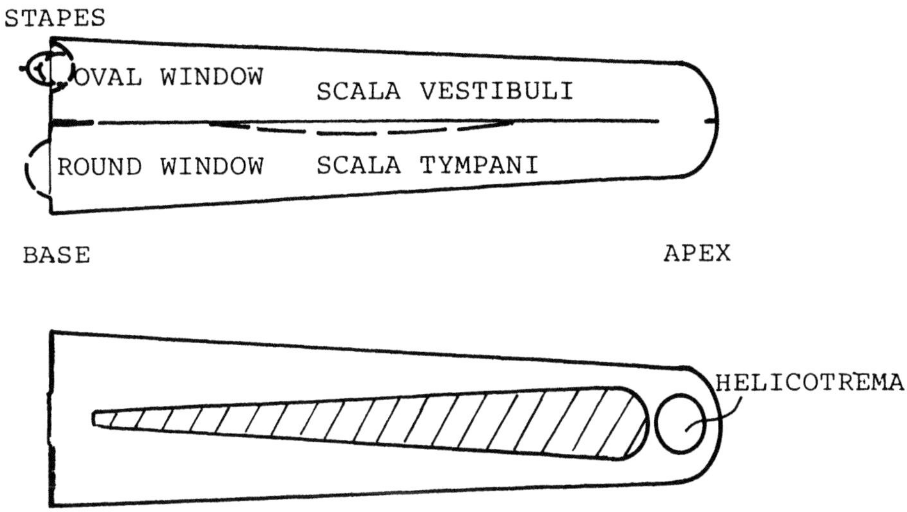

Fig. 12.2 Schematic representation of the uncoiled cochlea. The shaded area represents the basilar membrane which is the elastic part between the upper and the lower chambers (the scala media is neglected here). The basilar membrane is stiffer against the basal end but broadens (from 0.1 mm to 0.5mm) to the apical end.

When after a while of silence a tone starts with a compression the stapes will press the oval window membrane inside (broken lines). As a consequence, part of the perilymphatic fluid will escape by the helicotrema. The higher pressure in the scala vestibuli bends the basilar membrane downwards. The broad widening (broken line) sharpens with time when sound is presented periodically at constant frequency.

a maximal level of sounds in the mid-high frequencies only about 15 dB at the eardrum compared with the field. But the resonance characteristics depend also on the angle betweeen the acoustic source and the head, thereby helping to locate the origin of sound (SHAW, 1974).

In the middle ear the *ear ossicles (malleus (hammer), incus (anvil)* and *stapes (stirrup))* work as a system of levers to overcome the impedance mismatch between the air and the fluids of the cochlea. Without amplifying the force of the low-pressure sound waves in the air, the inner ear fluids would not be moved enough thus we hear at a similar level to when we have the head under the water. WEVER & LAWRENCE (1954) have demon-

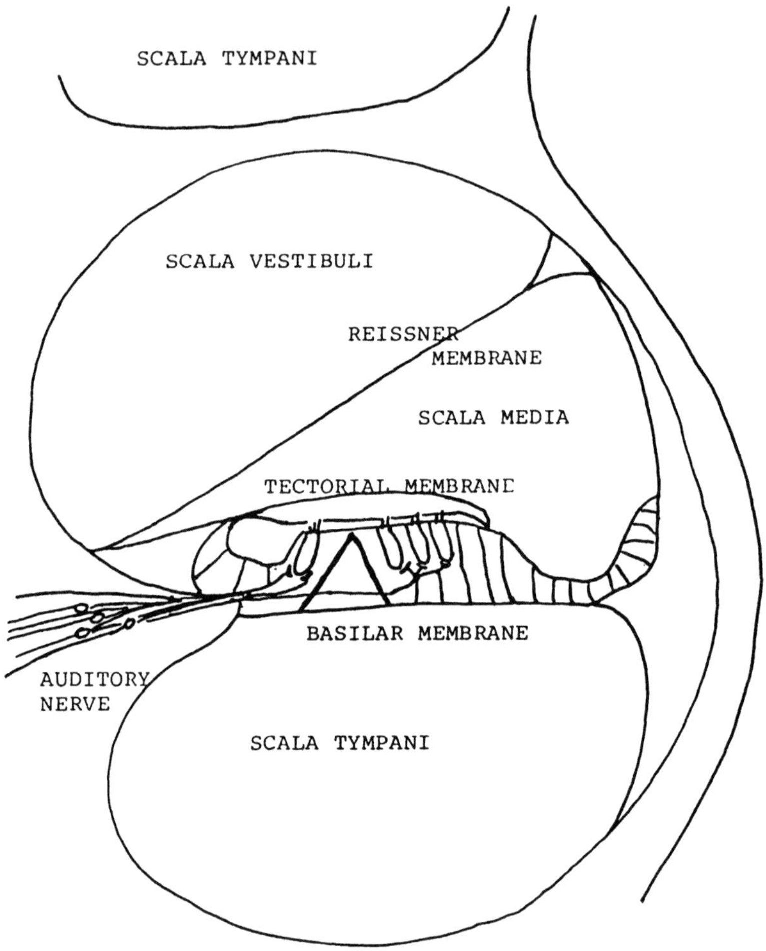

Fig. 12.3 *Schematic cross section of the cochlea marked by A-A in Fig. 12.1. If the pressure in the scala vestibuli is higher than that in the scala tympani, the organ of Corti and other parts of the scala media move down due to the compliance of the basilar membrane. The resistance of Reisner's membrane is very low and can be neglected. The movement of the basilar membrane will be registered by the organ of Corti, which transforms the mechanical movement of the basilar membrane into neural signals mostly with the help of the inner hair cells. At the base of a inner hair cell several synapses are situated. Every of these synapses is connected with an axon of the acoustic nerve. Most of the axons of the inner hair cells send their information to the brain (afferent neurons). The three rows of outer hair cells receive their information mostly from the brain (efferent neurons).*

strated that the transformation from the *eardrum (tympanic membrane)* to the membrane of the *oval window* is quite linear, i.e., when (at a certain frequency) the eardrum amplitude is doubled the membrane of the oval window also vibrates with a doubled amplitude.

The inner ear consists of the *vestibular apparatus* (the organ for equilibrium) and the *cochlea.* The cochlea is about 35 mm long and forms a cone-shaped spiral with 2.75 turns. The central core of the cochlea is called *modiolus.* It is easier to visualize the cochlea by imagining the spiral uncoiled, as shown in Fig. 12.2. In this figure the base of the cochlea is on the left and the *apex* on the right. On the base there are two openings which are closed by membranes: the oval and the round window. The cochlea consists of three chambers: *scala vestibuli, scala media* and *scala tympani* (Fig. 12.3). The scala media is filled with a fluid called *endolymph* and separates the other two chambers which are filled with *perilymph.* Scala vestibuli and scala tympani are connected at the apical end by a whole called *helicotrema.*

We assume that, after an interval of silence, a tone starts with the compression of the air. Through the conductive mechanism of the middle ear, the stapes presses the membrane of the oval window to the inside of the cochlea. The pressure in the scala vestibuli then will increase and the perilymphatic fluid will press the basilar membrane downward and also attempt to escape through the helicotrema. Since the fluid is nearly incompressible, the membrane of the round window bends outwards.

The most important mechanical process in order to get a hearing perception is the movement of the basilar membrane. The hearing receptors, i.e., the hair cells are mounted on the basilar membrane, and when the hairs (*cilia, stereocilia*) of these cells are moved, the fibers of the acoustic nerve will be stimulated.

When a periodic sigal is applied, every part of the basilar membrane will also move periodically. If we assume that the basilar membrane is divided into several segments along the way from the base to the apex, the maximum amplitude of every segment will depend on both the frequency and the loudness of the source signal. Basilar membrane motion can be seen by stereoscopic observations (BÉKÉSY, 1960) and the velocity can be measured with the help of the MÖSSBAUER EFFECT (RHODE, 1971). Periodic acoustic stimuli produces travelling waves at the basilar membrane, which start at the base, find a maximum and then descend quickly at the apical side. The place of maximum vibration depends logarithmically on frequency, i.e., tones of 100 Hz resonate close to the apex. Eeach time the frequency is doubled the place of resonance moves at a constant length towards the base, which is reached at the highest audible frequencies. The sharp relation between frequency of sound and the distance of the basilar membrane resonance from the base is called tonotopic organization. High frequency tones will cause only the basal part of the basilar membrane to swing, whereas low frequency signals will reach apical regions. When a combined tone is applied, the membrane will vibrate at different locations with high amplitudes, according to the frequencies contained in the acoustic signal.

Neural coding

Spiral ganglion cells are the primary auditory neurons forming the auditory (acoustic) nerve. About 95 % of them are afferent bipolar cells, connecting the inner hair cells with the cochlear nucleus in the brainstem.

Measurements of basilar membrane motion, using laser intecterometry (KHANNA & LEONARD, 1982), have shown that the tuning curves of the auditory nerve have a similar frequency selectivity as the basilar membrane. Every afferent fiber coming from an inner hair cell will react with the higest sensitivity to a special frequency, called the characteristic frequency of this fiber (frequency with the lowest threshold).

On the top of a hair cell cilia are grouped into a bundle consting of 30 to 150 hairs, 0.1 to 0.9 μm in diameter, and graded in length up to 7.2 μm in the base. (At the apex the longest are only 4.2 μm.) A constant displacement of the cilia result in a constant change of the intracellular potential of the hair cell. HUDSPETH & COREY (1977) have stimulated hair cells of the isolated frog sacculus. They demonstrated that the change in membrane voltage depends on the direction of the distortion: deflection toward the kinocilium (the largest of the cilia) causes depolarization (excitation), whereas the opposite deflection produces hyperpolarization (Fig. 12.4). Stimuli at right angles to the bundle's axis of mirror symmetry produce no change from the resting potential.

Although displacement-response relations vary from cell to cell and species to species, their constant steepness is characteristic: about 90% of the response range corresponds to a deflexion of only 50-120nm at the tip of the cilia (about 1°). In ordinary use, the hair movea distance less than their diameter (HUDSPETH, 1989). The threshold of hearing occurs at a deflection near 0.3nm (0.003^o), producing a 0.1mV increase of the receptor potential (CRAWFORD & FETTIPLACE, 1985), which evidently suffices for reliable synaptic transmission. In the hair cells the stimulating ionic channels are situated at the top of the cilia, but they are few in number, e.g., it seems that only one active transduction channel per cilium exists in bullfrog's sacculus (HUDSPETH, 1989).

Depolarization of the hair cell causes the generation of AP's in the primary fibers of the acoustic nerve. As a consequence of the directional excitation of the hair cell the firing times in these axons are synchronized with the acoustic signal*.

The average inner hair cell is innervated by at least 20 afferent fibers of the acoustic nerve (SPOENDELIN, 1969). All these axons generate spikes, even when no stimulating signal is applied. This *spontaneous activity* occurs to varying degrees in different fibers, but it disappears when the hair cells are lost (KIANG *et al.*, 1976; KIANG *et al.*, 1965). Thus, we can assume that spontaneous activity is caused by leakage or random release of neurotransmitter from hair cells.

Fibers with high spontaneous rates (50-60 *spikes/sec*) possess the lowest thresholds to acoustic stimuli. In spite of this high nervous activity we obtain no perception as long as the firing times are randomly distributed. Weak stimuli, below 3 to 4 kHz, will redistribute the firing pattern of the nerve array without an

---

* There is some contradiction concerning which direction of basilar membrane deflection the AP's in the acoustic nerve fibers are synchronized. The shorter latency of rarefraction clicks compared to compression clicks support the classical findings that deflection of the basilar membrane towards scala vestibuli excite the hair cell. This is in agreement with the conception that the upward moving basilar membrane displaces the sensory hairs away from the modiolus and towards the longest row of stereocilia (Javel, 1986). On the other hand experiments of Konishi & Nielson (1978) and Sokolich *et al.* (1976) suggest that the axons tend to have increased discharges with scala tympani motions.

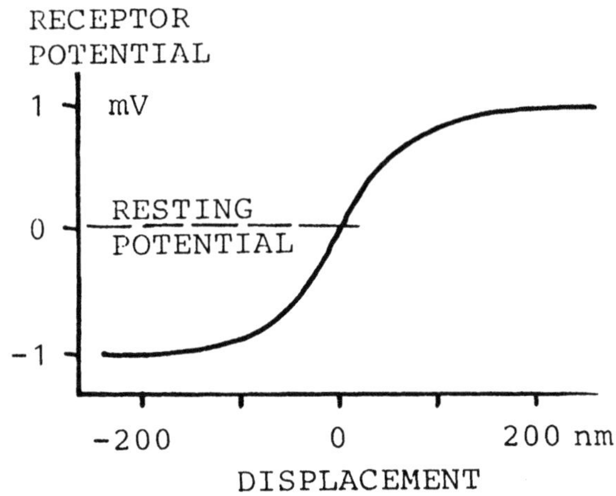

Fig. 12.4 Receptor potential as a function of cilia displacement. The sigmoid shape is typical, however, differences in the response are found. For example, receptor potentials in guinea pig inner hair cells yield voltage responses of 20-30 mV (Russell and Sellick, 1983). These inner hair cell responses are asymmetric with 4 times higher excursions to the depolarizing rarefaction peaks compared to condensations. See also Hudspeth and Corey, 1977.

(After Hudspeth, 1989)

increase in the total firing rate. The brain can interpret it as audible sound.

The fibers of the acoustic nerve react typically with an increase of the firing rate when loudness is increased within a range of 40 *dB* before they saturate. The tuning curves (Fig. 12.5) describe in the way the threshold of a fiber depends on the frequency of stimulating sound*. Beside the sensitive fibers there exist others with the same characteristic frequency but up to 80 *dB* higher thresholds (LIBERMAN, 1978; LIBERMAN & OLIVER, 1984).

Since 1943 the individual firing behavior of acoustic nerves have been measured by the insertion of microelectrodes (GALAMBOS & DAVIS, 1943; TASAKI, 1954; KIANG *et al.*, 1965). The first

* The strength of acoustic signals is measured relative to a sound pressure p_0 of 20 $\mu Pascal$ ($1\ Pascal = 1\ N/m^2$), which is nearly equal to the human hearing threshold at 2 *kHz*. The sound pressure level L is usually scaled logarithmically in decibels. $L=20 \cdot log(p/p_0)$ [dB], i.e., a 20 *dB* tone has the tenfold p_0 pressure.

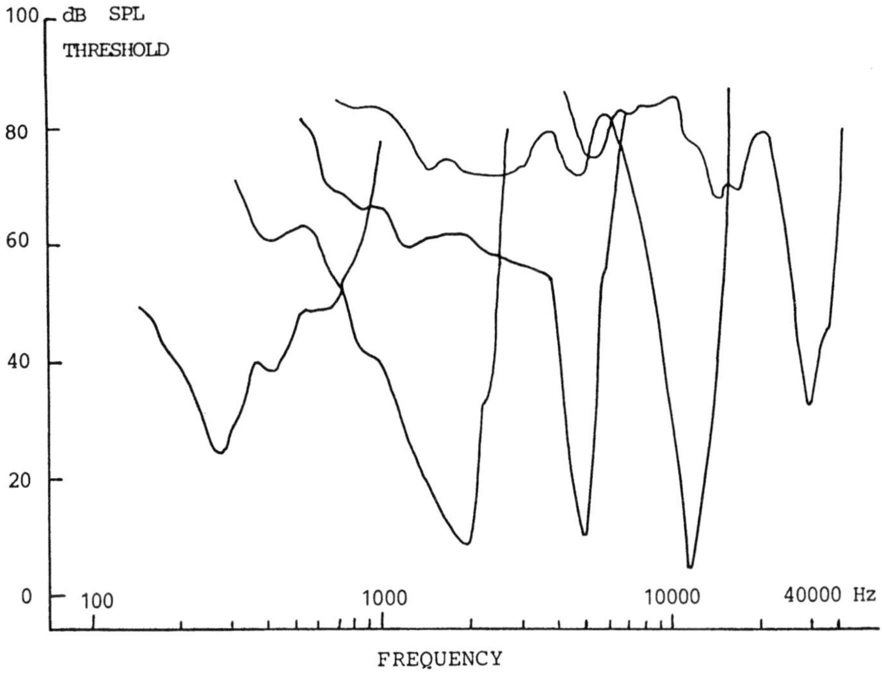

Fig. 12.5 Tuning curves of five auditory nerve axons of a single cat. The tuning curves of a single fiber describes the threshold sound pressure level (SPL). Typically, the tuning curves have a sharp minimum at the characteristic frequency, it broadens against the low frequency and becomes steep on the other side.

(After Javel, 1986)

AP elicited by a tone occurs 1-3 *ms* after the pressure wave strikes the airdrum. This delay depends on tone frequency (higher tones have shorter latencies). It is caused by the mechanical conduction, the synaptic delay at the hair cell, and the time required to conduct the AP to the recording electrode.

After the initial spike, the average fiber can respond in the first 10 *ms* with a high firing rate which can attain up to 2000 spikes per second. The rate then declines within the next 30 *ms* to a maximum steady state value of 150 to 300 AP's/sec. This discharge rate will decrease somewhat with time (JAVEL, 1986).

The higher centers of the nervous system obtain the essential frequency and loudness information by the simultaneous firing patterns of the individual fibers. Although the normal discharge rates of single fibers are usually lower than the frequency of the

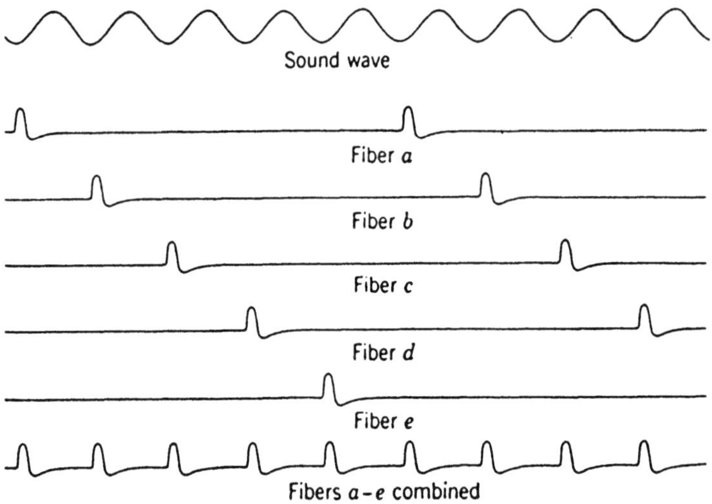

Fig. 12.6 *The volley principle. A periodic acoustic stimulus will cause firing patterns in different axons of one hair cell as marked with a - e. The combined signal contains all the minima of sound even when no single fiber is able to fire with this high frequency.*

(After Wever, 1949)

acoustic signal, the high innervation of inner hair cells let groups of fibers cooperate to represent the whole phase synchronized signal up to frequencies of 4-5 kHz.

In order to understand the cooperating effect, we assume that a tone of 1000 Hz will stimulate all the 20 fibers coming from a special inner hair cell, and each of them will fire with, e.g., 200 spikes/sec. If we subtract the delay, the firing times gather close to the minimum pressure times of the sound signal. We assume to find five fibers among them, which will react with the temporal behavior as shown in Fig. 12.6. Thereby, the entire riquired information about the minima of the stimulus is contained in the compound signal. This is the VOLLEY PRICIPLE introduced by WEVER (1949). If we assume that all the fibers of a hair cell would have the same sensitivity, we can obtain a good impression about the compound signal. This is possible even when we observe only one single fiber by the insertion of a microelectrode. When 20 equal experiments are performed sequentially, we can get an image of the firing pattern of the parallel process.

The firing times of a nerve are random in nature and there-

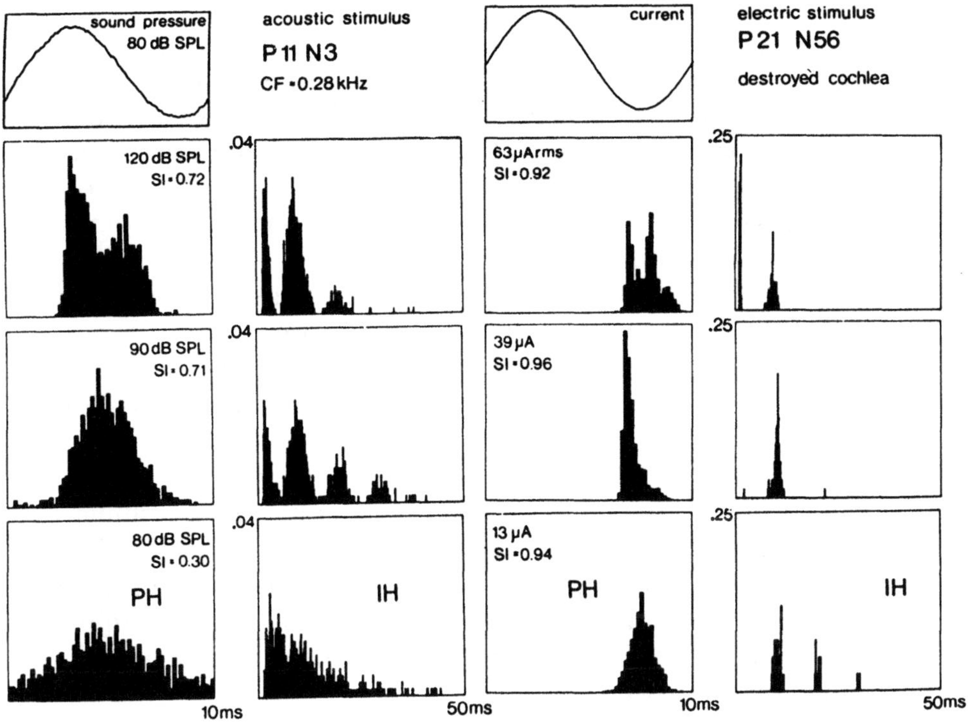

Fig. 12.7 Histograms obtained from single auditory nerve recording at sinusoidal stimulation.

Left: 100 Hz acoustic stimulation. As a consequence of the spontaneous rate weak signals (not shown) produce in period histograms (PH) an uniformly distributed noisy response, which gradually reflects the shape of the (decompressing) half wave of the stimulus signal when sound is presented with higher strength. When very high sound pressure is applied at low frequencies, the period histogram bifurcates as a consequence of double firing within a half period of the stimulus.

Right: 100 Hz extracochlear electrical stimulation of a deaf cat at the round window. The sharp responses and shorter interval histograms (IH) indicate better synchronism (no spontaneous activity) and higher firing rates compared to the acoustic stimulation. The interval histogram shows that when strong electrode current (63 μA) is applied every negative halfwave of the stimulus signal will produce at least one AP. The first sharp peak reflects the timing of multiple firing: double or even triple firing occurs with nearly the same interspike times.

(After Hartmann and Klinke, 1990)

fore a good overview on the firing pattern is obtained by using histograms. Fig. 12.7 shows period - histograms (poststimulus time histograms) which measure the relative latency between the minimal values of a sinusoidal stimulus signal and the arrival of the evoked spikes at the recording electrode*. In a similar way, interval histograms record the interspike times and give information on the missed minima of the source signal.

When stimulated with periodic acoustic signals at middle frequences the essential features seen in histograms are:

i) Very weak signals produce no significant histograms. They are close to the uniform distribution resulting from spontaneous activity.
ii) The shapes of the period histogram, as well as, the single parts of the interval histogram, are similar to the stimulating half wave of the acoustic signal.
iii) Stronger stimulation leads to higher firing rates and to better synchronization, while the number of missed minima is reduced. The period histogram becomes narrower and the interval histogram shorter.

Electrical stimulation of the auditory nerve

Although the general behavior of a single fiber is similar, when stimulated electrically (Fig. 12.7), there are some important differences compared to acoustical stimulations:

i) The firing rates are higher. This is demonstrated in Fig. 12.7 by shorter interval histograms, and especially by the early first peak in the case when stimulated with strong currents. Fig. 9.7 shows the short latencies between multiple AP's from single fiber recordings.
ii) There is less deviation in the firing times: the histograms are sharper.

---

* In order to obtain the period histogram of Fig. 12.7, the period of the 100 *Hz* signal was divided into 64 segments, each 10/64 *ms* long. The experiment lasted until 1000 periods occured. Every AP takes one place, in one of the 64 segments, and raises the corresponding bin by a constant amount. Thus, the histogram gives a picture of the relative probability of the occurence of a spike.

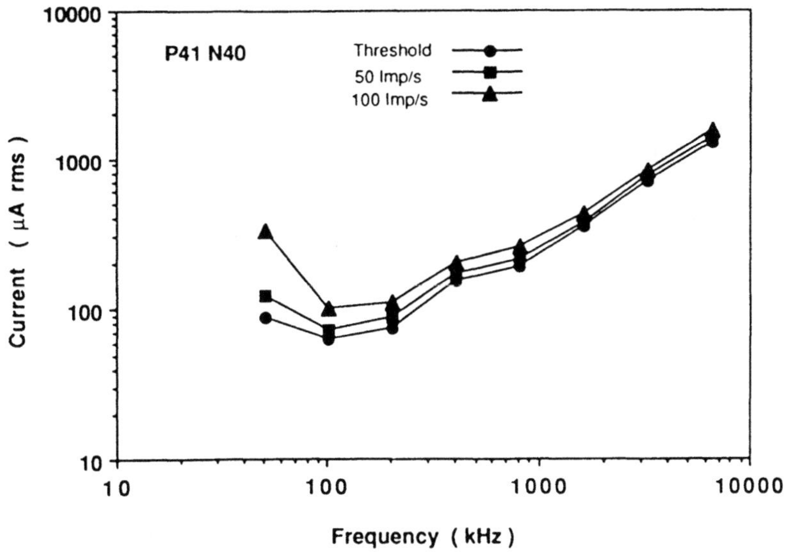

Fig. 12.8 *Isorate contours from a single auditory fiber of an acutely deafened cat obtained from sinusoidal round window stimulation.*

(After Hartmann and Klinke, 1990)

iii) The dynamic range is much smaller. (See also Fig. 12.8.)

In undeafened animals the neural responses are not as sharp as seen in Fig. 12.7 because the stimulation might produce mechanical vibrations causing additional neural activity (Fig. 12.9).

Fig. 12.10 shows the high synchronization obtained by round window stimulation with pairs of biphasic pulses in an acutely deafened cat. In these experiments of HARTMANN & KLINKE (1990) every superthreshold current pulse evoked an AP if the interval between two pulses was longer than 2 *ms*. With a delay of 1.3 *ms* only 76 % of the second pulse were followed by an AP, and with a 1 *ms* delay no second AP were produced. Multiple firing at low frequency sinusoidal stimulation (Fig. 9.7) show also an interspike time of 1.4 *ms*, similar to the shortest interspike times obtained from the double impulse experiment. Note the higher synchronization of pulsatile signals compared with sinusoidal stimuli.

The relative good agreement between the normal acoustic stimulation and the artificially evoked AP's allows the medical use of cochlear implants (Box 12.1). The best patients can under-

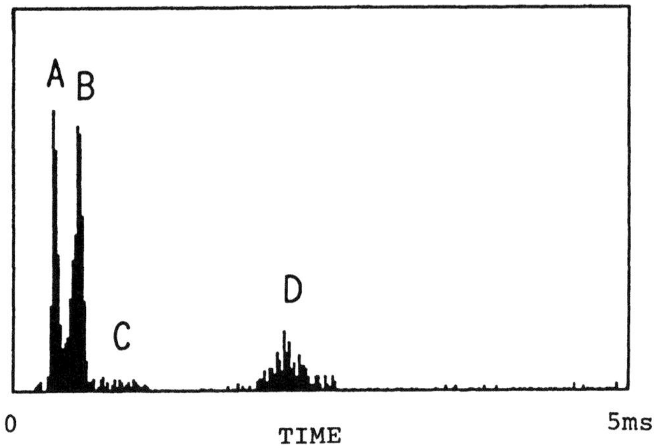

Fig. 12.9 *Period histogram obtained from undeafened cochlea stimulation with biphasic electrical impulses. Four different responses are seen:*

A *Latency: 0.3-0.5 ms. In undeafened animals, it is observed predominantly at high intensities, in deafened animals it is the main response already observed near threshold.*

B *Latency: 0.5-1 ms. This response seems to be caused by the second (negative) pulse.*

C *Latency: 0.8-1.2 ms. It is usually relatively weak, exists principally at intensities close to threshold.*

D *Latency: greater 1.5 ms. It has a low threshold, a wide dynamic range, and it is indistinguishable from the reaction generated by an acoustic click. The* D *response is absent in deafened animals. It is assumed (Moxon, 1971) that electric stimuli causes the traveling waves in the basilar membrane. That is, the electrical signal repels or attracts the negatively charged basilar membrane producing a click-like response, which is analyzed by the organ of Corti in the usual manner.*

Pulse duration: 200 μs/phase, repetion rate 200 Hz. The leading positive pulse is immediately followed by a negative one.

(After Javel, 1990)

stand about 90 % of words and sentences of an open list without the help of lipreading. Unfortunately, such good results are an exception and the average patient will use the implant for helping in lipreading. But, even in the outstanding cases, there remains a remarkable loss on perceptual quality. The main difficulty is to

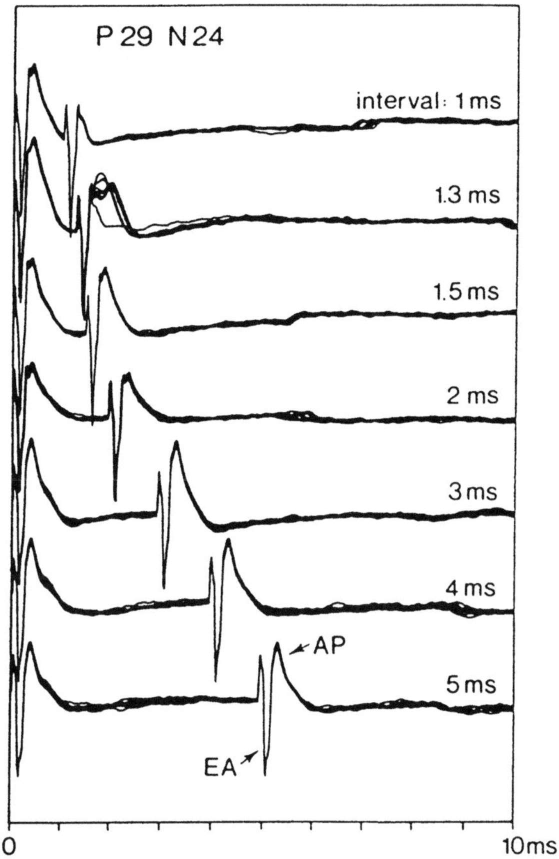

Fig. 12.10 Fiber responses when stimulated with pairs of biphasic pulses. Each row shows 10 original recordings superimposed with an electrical artifact (EA) and evoked AP. Stimulation of an acutely deafened cat via a monopolar round window electrode. Biphasic pulses of 100 μs/phase, starting with the cathodic signal.

(From Hartmann and Klinke, 1990)

produce an individual firing pattern in the auditory nerve which can be decoded by the auditory pathway.

Surprisingly, speech understanding of deaf patients which use extracochlear electrodes or single channel stimulation is often as good as that of the more sophisticated multichannel implants. This is caused by channel interactions during parallel stimulation, which reduce the advantages of place coding strategies. Before going into detail on the problems with multichannel applications,

BOX 12.1 **EAR PROSTHESES**

Ear prostheses for completely deaf people have been developed by various teams. Two types of implants are commonly used: extracochlear implants have mostly a single active electrode fixed in the round window niche, intracochlear implants are usually introduced via the round window into the scala tympani where several electrodes allow multi-channel stimulation. Several types of bipolar electrodes are in use for multichannel stimulation. These are either pure radial, pure longitudinal, radial with offset, or banded longitudinal (see Fig. 12.27).

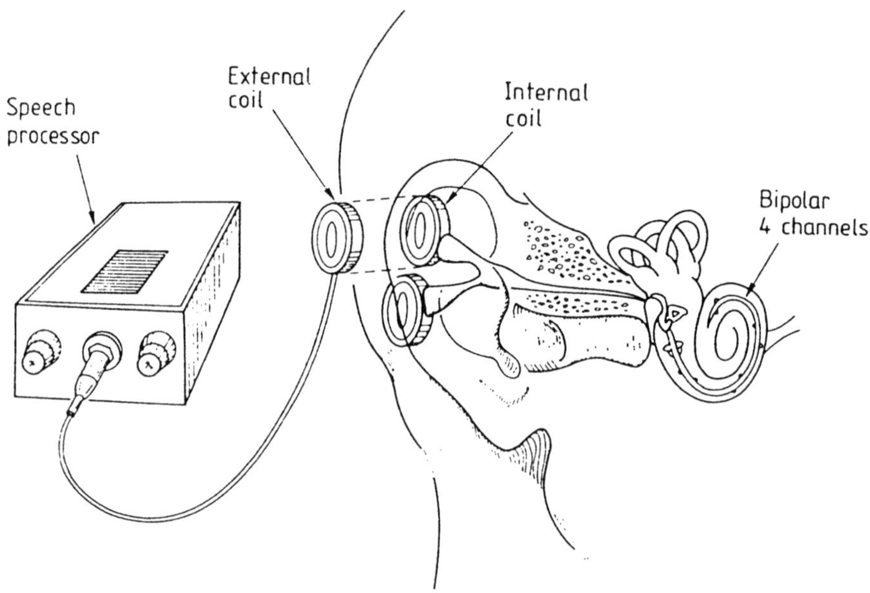

The basic components of a typical cochlear implant. The speech processor compresses the amplitudes of the acoustic signal and supports the external coil which sends the signal transcutaneous via internal coils to the electrode. The transmission is usually inductive and needs no extra power supply at the implanted side. The signals at the electrodes are either analog or pulsatile.

(After Hochmair et al., 1984)

we will be concerned with the time responses obtained from a single active electrode.

Simulation of the nerve array response by local models

We will see below, that it is difficult to model the complicated geometry of the inner ear in detail. Neverthless, we can obtain a first approach on the firing pattern by neglecting spatial influences. We have noticed, that short pulsatile signals will produce unnatural hypersynchronized neural responses. The question is, how does the whole nerve answer depend on the shape of other forms of signals, e.g., when stimulated directly with speech signals from a microphone.

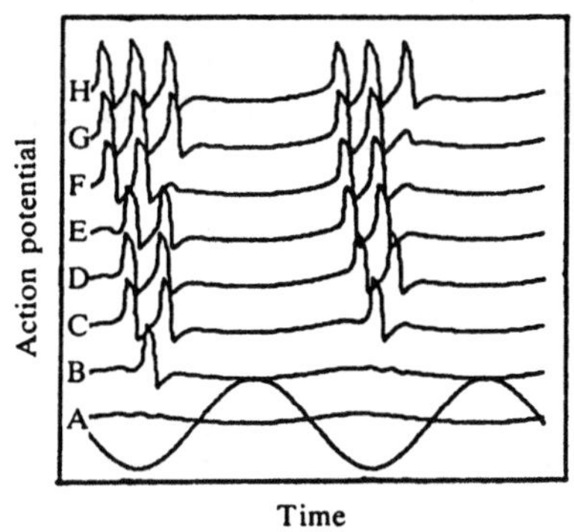

Fig. 12.11 *Simulated nerve response to a sinusoidal signal at 100 Hz. Stimulus strength was increased linearly from the lowest, trace* A, *(subthreshold response) to the highest, trace* H, *where triple firing occurs within the stimulating half period. Trace* A *corresponds to a distant fiber, traces* B *to* H *correspond to fibers at decreasing distances from the electrode.*

Simulation using standard Fitzhugh model without noise (Box 5.1). The time scale with $\beta = 7$ was adjusted to the experimental 1.4 ms interspike interval found in Fig. 9.7.

(From Rattay and Motz, 1987)

Even with the simple local FITZHUGH MODEL of Box 5.1 it is possible to predict some essential features of the neural reactions. An example is given in Fig. 12.11 which demonstrates the multiple shooting phenomenon as described in the experiments of

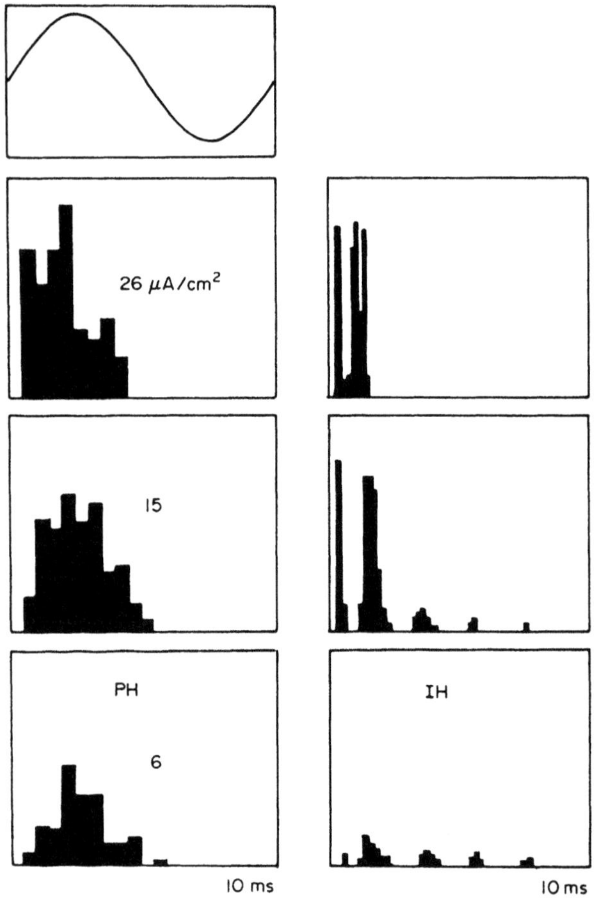

Fig. 12.12 *Simulated fiber responses for sinusoidal electrical stimuli. Period histograms (left) and interval histograms (right) at different stimulus strengths show a similar behavior as found from experiments. The first high peaks in IH, at 15 and 26 $\mu A/cm^2$, indicate multiple firing within a half period.*

Stimulation was with standard Hodgkin Huxley data, but k=12 (Eqn. HH-5 of Box 4.1). Note that the current signal becomes stimulating in the local model when it is positive, but the current should be inverted for interpretation of cathodic stimulation with extracellular electrodes.

(After Motz and Rattay, 1986)

HARTMANN et al. (1984) for higher stimuli strengths (Fig. 9.7). This model was used without a noise term; therefore, there is no variance in the response times for fixed stimulius strength. By introduction of a small noise current, local models can also be used to reproduce a similar variance as known from experiments. This is seen by the comparison of Fig. 12.12 and 12.7.

Of all the nerve models described in Chapter 4 and 5, only that of HODGKIN & HUXLEY and that of FITZHUGH are suitable for auditory nerve simulation concerning the reactions with low frequency stimulation. Only these models will react with multiple firing (Fig. 12.11 and Fig. 12.12), which may reflect a phenomenon that was described as the hearing of a ringing sound when low frequency signals were presented to some patients with cochlear implants.

The temporal behavior of the models has to be adjusted to experimental knowledge by the time transformation coefficient ($\beta = 7$ in BVF model) or by the advance of the thermic acceleration factor k for the gating mechanism. In the HH model the time, separating two AP's in the case of double firing, is 1.6 ms for k=9, 1.5 for k=10 and 1.4 for k=12. From the experiments described in Fig. 9.7 it is seen that k=12 may be the most appropriate value.

It is surprising that the HH model should be used to simulate the time pattern in the auditory nerve; although, it is assumed that the spiral ganglion becomes excited in the myelinated part. None of the other models will react to constant current stimulation with multiple firing. Therefore, these models are not able to reflect the double shooting phenomenon observed at low frequencies. Another discrepancy reguards the recovery time of the models. The CRRSS- and SE MODEL, as well as, the modified FH model will react with AP's suited in their time shape, but their refractory behaviour is only suited to describe motor nerve reactions. In contrast to the motor nerves, which have chronaxies that cluster at 100 μs (LOEB *et al.*, 1983), the chronaxie of the auditory nerve is about 350 μs (COLOMBO & PARKINS, 1987) which is very close to the value of the HH model (340 μs for k =12). This was illustrated by the strength duration curve in Fig. 4.8.

The better applicability of the HH model was further demonstrated by a double pulse experiment performed by DILLIER (private communication): A signal consisting of two 100 μs pulses with variable strengths I_1 and I_2, separated by 100 μs according to Fig. 12.13, was used for stimulation. First $I_2 = 0$ (single pulse experiment) and the range of perception can be explored by varying I_1. Then I_1 is held at a fixed value well below the upper limit

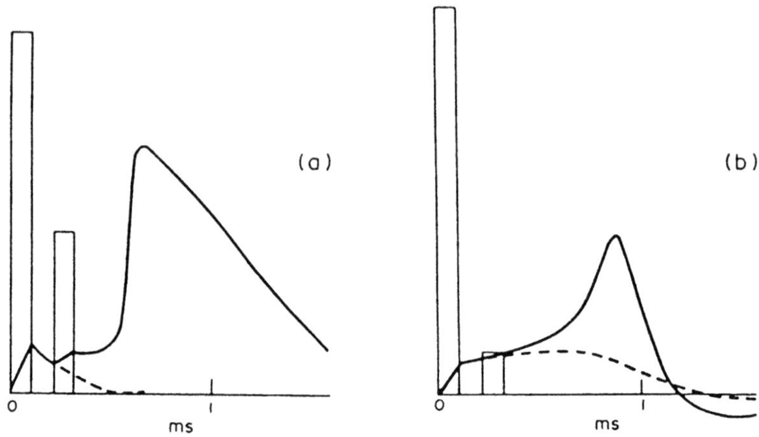

Fig. 12.13 Double pulse threshold experiment simulated with the FH model (a) and the HH model (b). The fiber is stimulated by a first pulse 3% below threshold. The second pulse is adjusted to produce just a superthreshold response (full line). If the second pulse is missed (broken line), the FH model decreases quickly whereas the HH model predicts a maximum membrane voltage more than 0.5 ms after the first pulse is switched off. A considerable lower second pulse is necessary to generate an AP when simulated with the HH model. The second impulse needs 46 % of I_1 in case (a) and only 10 % in case (b).

If (a) would be calculated with the modified FH equations the reactions become quicker. This require even a stronger second pulse and causes greater difference to experimental results.

Pulse duration and delay 100 μs. Simulation is with standard FH model data (37° C, producing slow AP's) and standard HH data, but with k=12.

(After Motz and Rattay, 1986)

of the range. The double pulse is offered alternatively with the single pulse and the patient is asked whether he can distinguish the two signals. In the just distinguishable case, the relation between I_1 and I_2 was found to be very similar to that resulting from a simulated threshold experiment, described in Fig. 12.13. We will interpret the just distinguishable case by inspection of the whole nerve array answer: In the single pulse experiment a certain population of fibers will be excited. If the second pulse leads to another perception, additional fibers, more distant from the electrode, will also produce spikes. These fibers would react

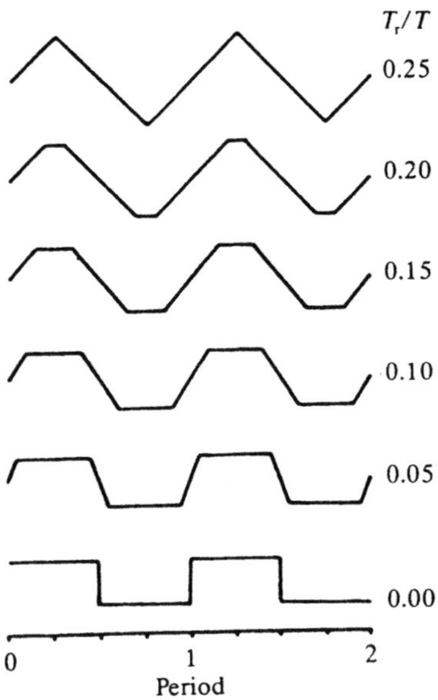

Fig. 12.14 Signals with varying rise time per period (T_r/T) used by Dobie and Dillier (1985). The amplitude of the electrode current is normalized to deliver a unit charge per half period for each case.

in a subthreshold manner by the single pulse experiment, but nevertheless, their membrane potential arising from the first pulse is remarkable. Nerve models predict different behavior within the interpulse time. Even the non-modified FH model, which reacts relatively slow, decreases rapidly when the first pulse is switched off. In contrast, the HH model (modified to react quickly by using k=12) has a small increase in the interpulse interval.

Now, we will illuminate the multi-channel activity of the auditory nerve by analyzing another experiment perfomed by DOBIE & DILLIER (1985). They examined the ability of cochlear implant patients to discern the different waveforms varying according to Fig. 12.14 from rectangular over trapezoidal to triangular shapes. The signals are characterized by the rise time per period (T_r/T) which varies from 0 (rectangular signals) to 0.25 (triangular signals). Their experimental results show that the waveforms can be discriminated even in 1kHz signals (Table 12.1).

When stimulated electrically, the AP's will occur in the negative half wave of the signal. Fig. 12.15 shows remarkable delays for distant fibers in the firing pattern obtained within the negative half wave of a triangular shaped stimulus at 250 Hz. At

Table 12.1
Waveform discrimination by normal subjects and patients fitted with single-channel cochlear implants.

Frequency [Hz]	Difference limen of rise time [ms]		
	normal subject	patient UT	patient ET
80	0.12	0.88	—
125	0.08	0.96	0.88
250	0.04	0.08	0.56
500	0.04	0.06	0.34
750	0.04	0.18	0.02
1000	0.02	0.17	0.15

(After Dobie and Dillier, 1985)

low frequency stimulation muliple firing occurs with differences in the firing pattern depending on signal shape (Fig. 12.16). It seems that the multiple shooting patterns are confusing. This may explain why the performance falls off sharply at low frequencies. The performance also deteriorates at high frequencies. This can be seen from Fig. 12.17 where the standard deviation of the simulated whole nerve response is low.

We assume that the central auditory nervous system will compare the firing times in different fibers in order to find differences of the presented waveshapes. We are looking for intrinsic characteristics of the discernible firing pattern presented in Fig. 12.18 since reference signal and trapezoid waveforms are offered at different times. Maximum timing differences between neighboring fibers occur at the most distant stimulated axons. Maximum time differences within a distant fiber bundle at 250 Hz are 260 μs (marked by arrows in Fig. 12.18b) for the reference signal and 190 μs for the discernable trapezoidal waveform. The difference between these times, namely 70 μs is the clue. At 500 Hz these times are respectively 160 μs, 90 μs and 70 μs (Fig. 12.18c). Since distant fibers are more numerous, we shall focus attention on them. At 1000 Hz we may take the difference between the standard devia-

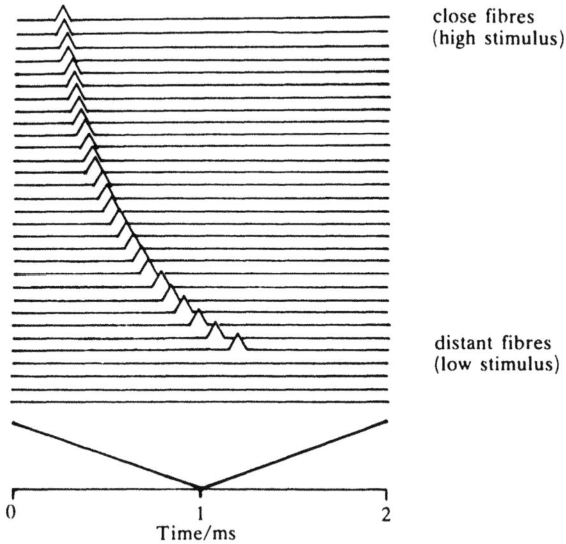

Fig. 12.15 *Simulated timing of AP's evoked with triangular signal of 250 Hz. Horizontal lines correspond to fibers excited by linearly increased stimulus strength. Firing times are indicated by markers. Simulation with the BVF model, standard data, $\beta = 7$.*

(After Rattay and Motz, 1987)

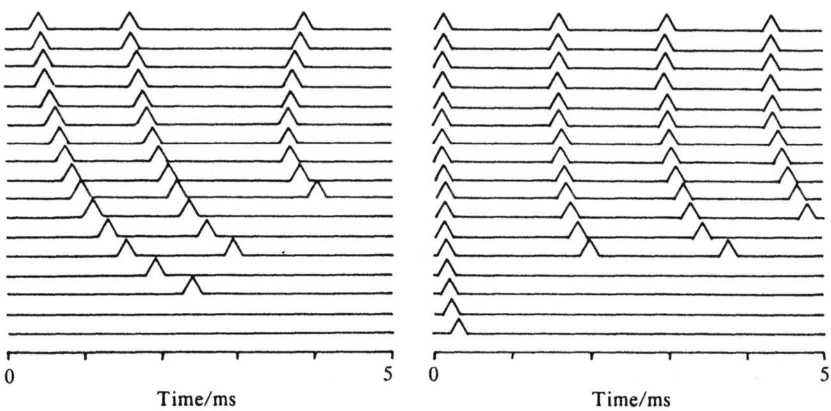

Fig. 12.16 *Simulated firing patterns for nerves stimulated with triangular (left) and rectangular (right) waveforms at 100 Hz. Simulation with the BVF model, standard data, $\beta = 7$.*

(After Rattay and Motz, 1987)

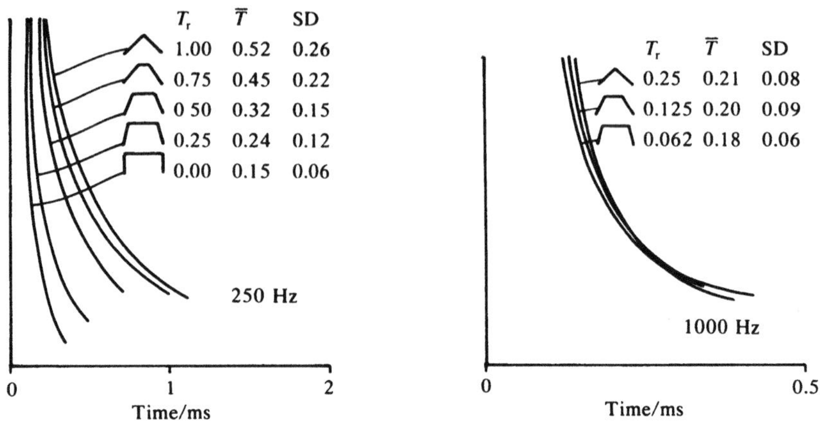

Fig. 12.17 *Simulated timing of AP's evoked by stimulation of the nerve with varied waveforms. Curves relate stimulus strength to the time of firing for the various shapes. The origin of the time axis marks the start of the negative stimulus halfwave. The rise time (T_r), mean time delay ($\bar{T}$), and standard deviation (SD) of delay are given. Simulation with the BVF model, standard data,* $\beta = 7$.

(After Rattay and Motz, 1987)

tion for the rectangular wave (68 μs) and that for the trapezoidal waveform (81 μs), which is 13 μs, as the clue. At 125 Hz it is possible that the clue is the difference between double and triple firing, whereas for the higher frequencies the clue is provided by very short time differences. However, we know that the central auditory nervous system is able to discriminate such short times, e.g., the interaural time difference limen for directional hearing is in the order of 10 μs (KLUMPP & EADY, 1956).

A single-channel signal processing strategy

We have seen in Fig. 12.8 that the auditory nerve can also be stimulated electrically by high frequency signals. The firing pattern at high frequency stimulation must be quite unnatural, because it cannot produce adequate sensations. Unfortunately, the essential information in human speech is carried by frequencies up to 2-3 kHz (Box 12.2). Better speech understanding would be

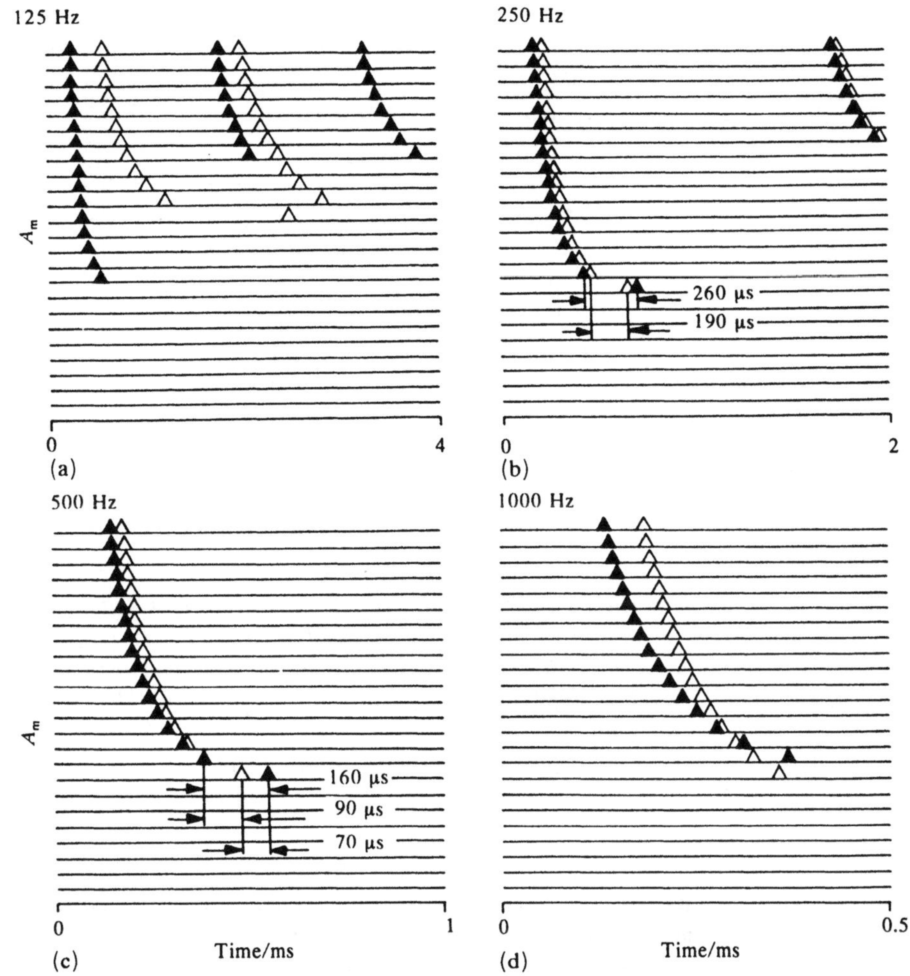

Fig. 12.18 Timing patterns for the difference limen of patient UT (Table 12.1). Full triangles mark the firing times when stimulated by rectangles, empty triangles mark the just discernible trapezoidal stimulus shape. At 125 Hz triple firing occurs for rectangular stimuli, but double firing occurs in the discernible trapez signals. The timing curves cross for 250 Hz, 500 Hz and 1000 Hz. Simulation with the BVF model, standard data, $\beta = 7$.

(After Rattay and Motz, 1987)

possible either by using sophisticated single channel strategies or by refined techniques with multi-channel electrodes, which would reduce the signal mixing produced by channel interactions.

The simplest way to apply speech presentation to a deaf sub-

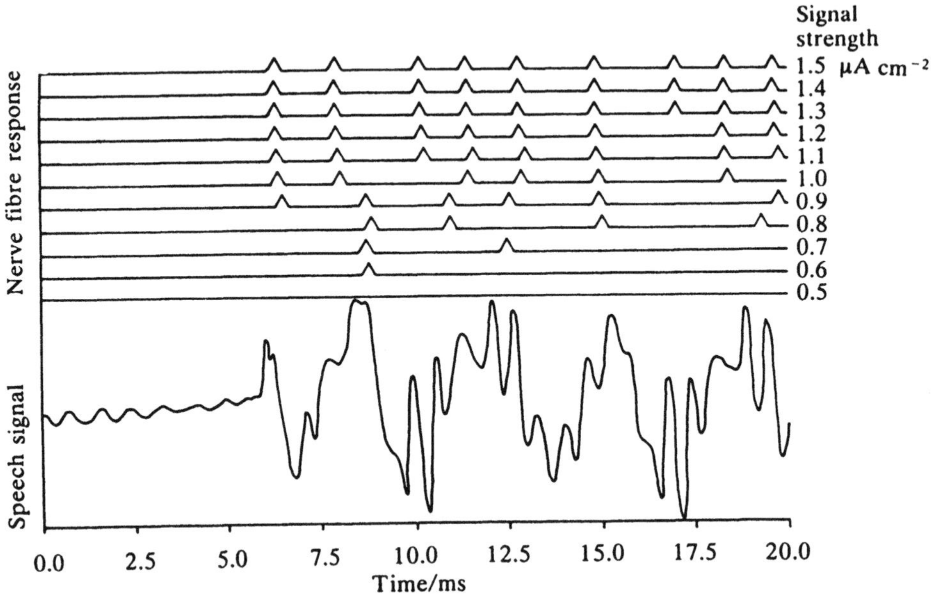

Fig. 12.19 *Simulated response to the German vowel /y:/. After a period of silence, the vowel is presented to a cochlear implant patient. The lower trace show the scalp potentials (reproduced with courtesy of v. Wallenberg) generated by the implanted electrode. The upper traces represent the firing times for individual fibers with different distances to the electrode. The upper traces correspond to high signal strength and reactions of fibers close to the electrode.*

To a normal listener, the scalp potential sounds natural when presented by a loudspeaker, this means no essential speech features are lost on the way to the stimulating electrode.

Simulation was with standard HH model, k=12, no additional noise. In the local model the input signal has a stimulating (polarizing) effect when it is positive.

(After Motz and Rattay, 1987)

ject via electrical stimulation is by compressing the amplitudes of the acoustic signal, thereby fitting the source signal to the small comfortable range allowed for electrode currents. These compressed signals are well understood by a normal listener, even when the degree of compression depends on frequency and, when a low pass filter is applied. Such compressed speech signals can be used to supply a stimulating electrode (positioned, e.g., near the round window) with an analogous current. A first approach of the evoked neural activity can be found by using this signal as

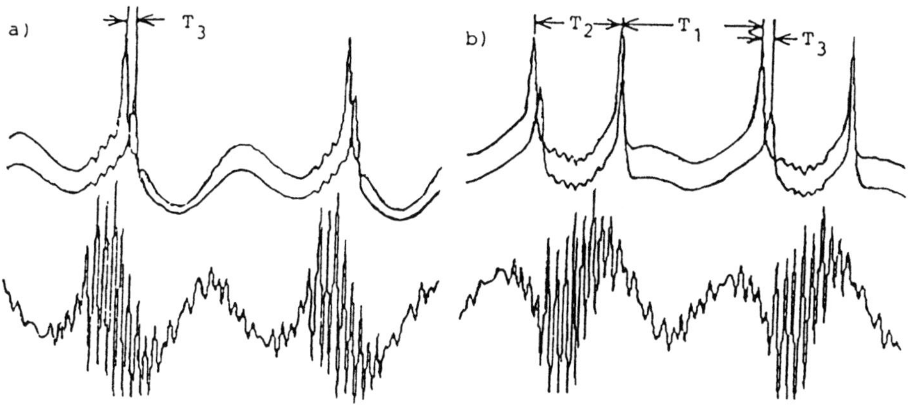

Fig. 12.20 Signal currents at the stimulating electrode and the resulting firing pattern from fibers in neighbored regions. The vowel /i:/ of a male voice (bottom curves) generates the firing patterns shown in the upper traces when presented as stimulus signal with the intensities 350 $\mu A/cm^2$ (top) and 250 $\mu A/cm^2$. Small time differences, that cannot be carried by a single fiber, are contained in the compound signal according to the Volley principle. In some cases the characteristic frequencies of the speech signal will be obtained via the firing time differences in neighboring fibers.

The firing patterns depend on polarity: In case (a) the times T_1 and T_2, which correspond to the first and second formant, are missed. But by accidentlally inverting the signal (case (b)), a firing pattern is obtained which includes all the main frequencies necessary to understand the speech signal.

10 ms of signals are presented. T_1=4.75ms, T_2=3.1ms, T_3=0.36ms. Simulation with standard HH data, k=12.

(After Motz and Rattay, 1986)

an input to a local nerve membrane model as described above.

Fig. 12.19 shows the simulated nerve array response to the stimulation of the German vowel /y:/. It can be seen that in the source signal one peak follows another rapidly, but no single fiber is able to represent this frequency information. Even the compound signal will not represent all the peaks of the speech signal which has been low-pass filtered. We must conclude that essential information is lost, although sometimes the higher frequencies generate a corresponding interspike time in neighbored fiber is able to represent this frequency information. Even the compound signal will not represent all the peaks of the speech signal which is already low-pass filtered. We must conclude that essential informations are lost although sometimes the higher frequencies generate a corresponding interspike time in neighbored fibers. [Compare, e.g, the firing times in Fig. 12.19 evoked by the signal strengths 0.9 and 1 $\mu A/cm^2$.]

Fig. 12.20 gives another example of simulated responses in neighbored fibers. In favorable cases the interspike times correspond to the essential speech features, namely to the basic speech frequency and to the first, second and third formant. Unfortunately these corresponding interspike times are poor, and as demonstrated in Fig. 12.20, the firing pattern depends on the polarity of the source signal. Change of polarity evokes other percepts in cochlear patients.

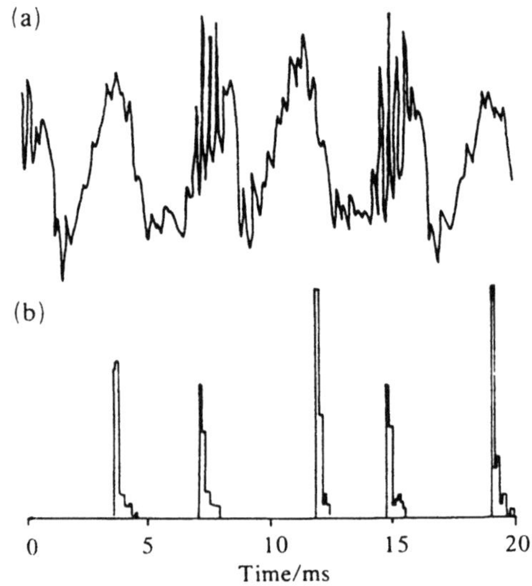

Fig. 12.21 Simulated nerve response to the vowel /i:/. The speech signal (a) produces a firing pattern which is strongly synchronized with the basal wave of the speech signal but higher frequencies are represented poorly as seen from histogram of firing times (b).

Simulation with standard HH model, k=12. 50 runs, with amplitudes varied from threshold to 5.5 times threshold.

(After Motz and Rattay, 1987)

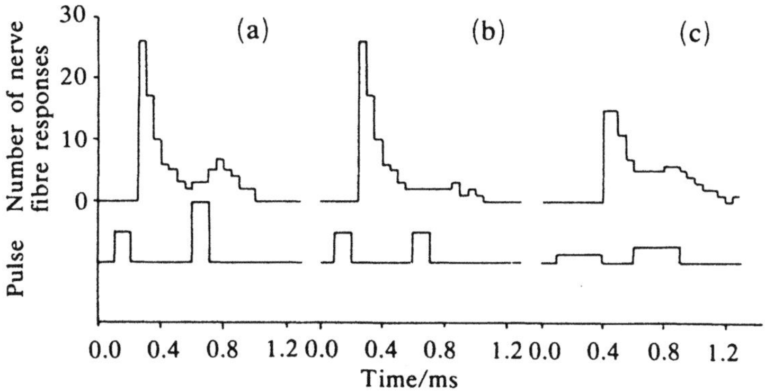

Fig. 12.22 *Nerve firing times by double pulse stimulation. The interpulse time of 0.5 ms is best represented in the fiber response, when the duration is short and when the second pulse is strong (case a). Thereby, a second population of more distant fibers is stimulated with a delay close to the interpulse time. Equal pulse amplitudes will only stimulate a small second population (case b). When longer stimuli are applied, the firing times are not sharply synchronized (case c).*

Simulation was with standard HH model, $k=12$. Pulse durations 100 μs in (a) and (b), 300 μs in (c). Pulse strengths $I_i=A_i \cdot z$ and z is varied from 100 to 500 in steps of $\Delta z=4$ (100 runs corresponding to 100 fibers with different distances to the elctrode); $A_1=0.5$, $A_2=1$ in case (a), $A_1=1$, $A_2=1$ in case (b), $A_1=0.16$, $A_2=0.32$ in case (c).

(After Motz and Rattay, 1987)

First, it is instructive to study the responses of fibers to several pulses with an interpulse interval of, e.g., 0.5 ms corresponding to the second formant of a speech signal. Fig. 12.22 illustrates the behavior by a simulated double pulse experiment. In order to find the second formant well represented in the poststimulus histogram, the second impulse should be stronger than the first and the pulse durations should be short.

For practical applications the monopolar pulses should be replaced by bipolar ones in order to avoid charge accumulation. Fig. 12.23 presents the theoretical behavior of the nerve response when stimulated with a sequence of 4 pulses using the strategy of increasing pulse strength.

From the foregoing consideration, it becomes obvious that an improvement in vowel identification might be obtained with a pulse strategy: Within a (negative) slow half wave of the speech

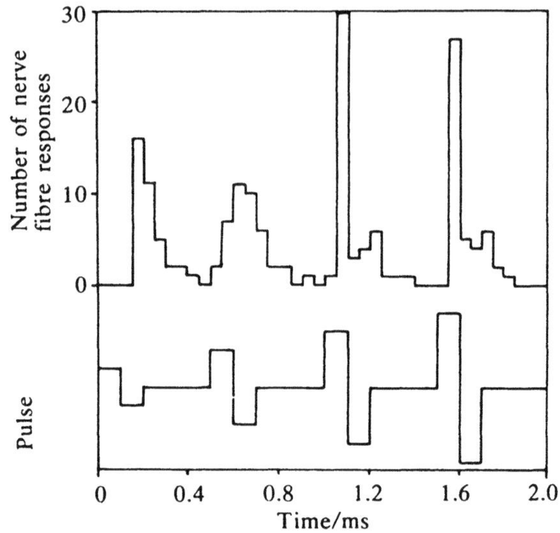

Fig. 12.23 *Nerve response to a pulse sequence with linear amplitude rise. The firing pattern of 100 axons representes the auditory nerve in a single electrode stimulation experiment. This is shown in the form of a poststimulus histogram (upper trace) which was a consequence of the stimulating signal (lower trace).*

Simulation was with standard HH data, k=12. Amplitudes of the biphasic pulses: $A_1=1$, $A_2=2$, $A_3=3$, $A_4=4$; *Pulse strengths* $I_i=A_i \cdot z$. *z is varied from 60 to 300 in steps of* $\Delta z=2.4$ *(100 runs corresponding to 100 fibers with different distances to the electrode).*

(After Motz and Rattay, 1987)

signal, the strategy can evoke distributed high frequency answers according to the natural volley principle in neighbored fiber regions. In the following (positive) halfwave nothing will be sent and then the procedure starts again. As an example, Fig. 12.24 illustrates that the poor neural spectrum can be improved in order to contain the essential speech features.

Multichannel electrodes

With the help of multichannel electrodes the natural place coding should be imitated: By using several bandpass filters, the frequencies of the source signal are separated. According to the natural distribution, high frequency signals are then presented at

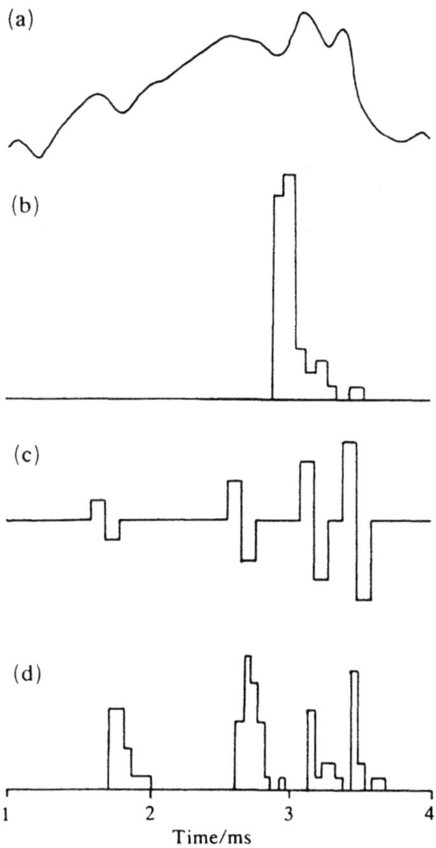

Fig. 12.24 *Comparison of normal neural answers and results of pulse strategy coding. (a) Half wave of speech signal /i:/, as presented at the cochlear electrode, and (b) the total nerve response in the form of a poststimulus histogram. (a) and (b) are details of Fig. 12.21. (c) Sequences of bipolar pulses corresponding to the maxima of the speech signal. (d) Histogram of the HH response for fifty runs.*

(After Motz and Rattay, 1987)

the basal end of the electrode array, whereas, the lower frequencies will be sent from more apical electrodes. Initial results using such electrodes have been disappointing because speech understanding is in the same order as the stimulation without frequency separation (HOCHMAIR-DESOYER *et al.*, 1981). This relatively bad performance is probably caused by a current spread along the scala tympani which has a large length constant of 3-12 *mm* (BLACK & CLARK, 1980).

Experiments on channel interactions (e.g. GLASS, 1985; VAN DEN HONERT & STYPULKOWSKI, 1987; HARTMANN & KLINKE, 1990) show that a place-dependent stimulation of the auditory nerve is possible on principle, but there are strong interactions during parallel stimulation if the distance of the channels are too small. Bipolar electrodes have sharper located influences on the target fibers than monopolar ones. The best focusing effect with bipolar electrodes is reached by using pure radial electrodes (Fig. 12.27).

Simulation of the auditory nerve response with spatial models

There are four essential differences in modeling the excitation of the auditory nerve compared to the excitation of motoneurons:

i) Much more attention has been paid to the selective firing times of different neurons in order to adjust the natural coding.

ii) The inner ear has a complex geometry with resistivities varying from 60 to 70 $\Omega.cm$ for the endolymphatic and perilymphatic fluids to 1800 $\Omega.cm$ (2 $k\Omega/mm$) for the basilar membrane and 60 $k\Omega.cm$ (27 $k\Omega/mm$) for the Reisner membrane. (FINLEY *et al.*, 1990; STRELIOFF, 1973)

iii) The geometry of the auditory nerve fibers differs from the simple rule for myelinated fibers that the internodal distance is about 100 times the outer fiber diameter.

iiii) The auditory nerve axon membrane reacts slower than that of motoneurons: AP's are somewhat longer, but the values for time constants and chronaxies (of about 350 μs) are remarkably longer than the value for a typical motoneuron (100 μs).

Fig. 12.25 shows schematically the anatomy of the primary auditory nerve fiber simplified after data of LIBERMAN & OLIVER (1984). The fiber is unmyelinated at an inner hair cell. After passing the *habenula perforata* it becomes myelinated and even the cell body is covered with multiple sheets of myeline. The internodal lengths of the first four segments are about 150 μm (after the third segment the fiber reaches the cell body). Within the next segments the internodal length increases about 50 μm per segment to a length of about 350 μm. This internodal length is more or less constant for the rest of the axon. There exist about 15 nodes of Ranvier before the axon passes the Schwann-glial border into the cochlear nucleus. The distal part of the axon is smaller in diameter than the central part. The ratio of the process diameters is correlated with the spontaneous firing rate. All neurons with ratios greater than 5 are of low or medium spontaneous rates (LIBERMAN & OLIVER, 1984).

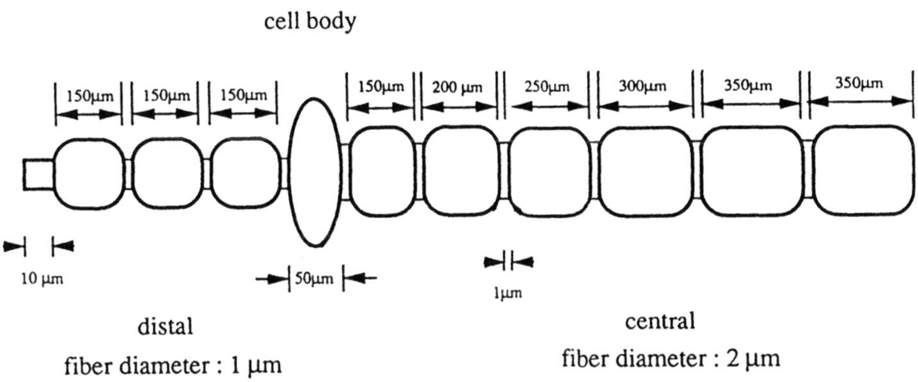

Fig. 12.25 Simplified morphometry of a typical primary auditory nerve fiber with low characteristic frequency (0.8 -3 kHz). The fiber starts with an unmyelinated part of about 10 μm in length; it has three myelinated segments between the inner hair cell and the cell body, and about 12 segments on the other side before entering the cochlear nucleus.

COLOMBO & PARKINS (1987) have used similar data of LIBERMAN & OLIVER in order to calculate the electrically stimulated auditory nerve response. They especially modified the internodal lengths in order to obtain with the FH model the chronaxies known from experimental evidence.

An outstanding three-dimensional model that take into consideration the complex geometry of the inner ear was developed by FINLEY *et al.* (1990). By using the finite element method they can compute the potential distribution evoked from arbitrary electrodes along the target neurons.

A 5.2 mm long section of the cochlea is parted into 12 layers with high resolution in the center according to Fig. 12.26(a). In a

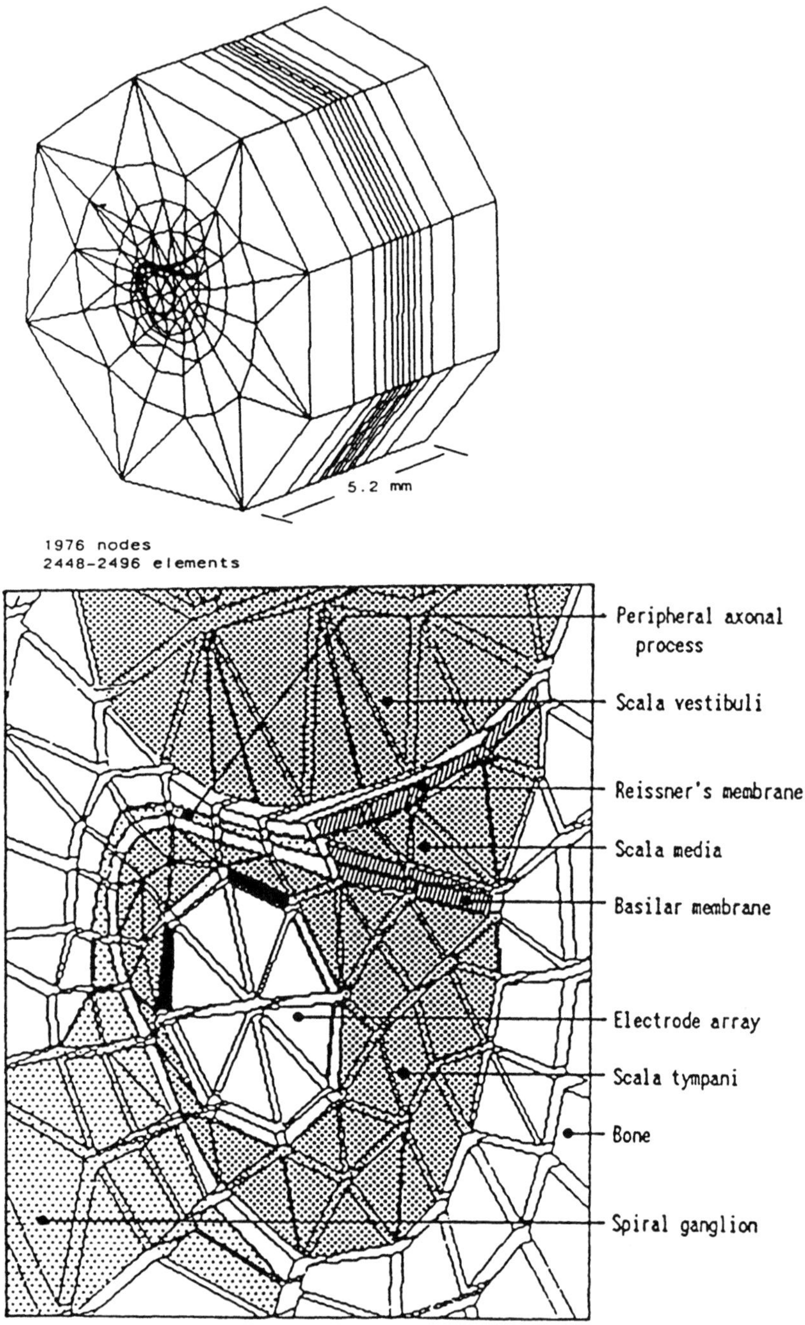

Fig. 12.26 *Three-dimensional model of a section of the uncoiled cochlea (a) and enlarged view of the central region (b).*

(From Finley et al., 1990)

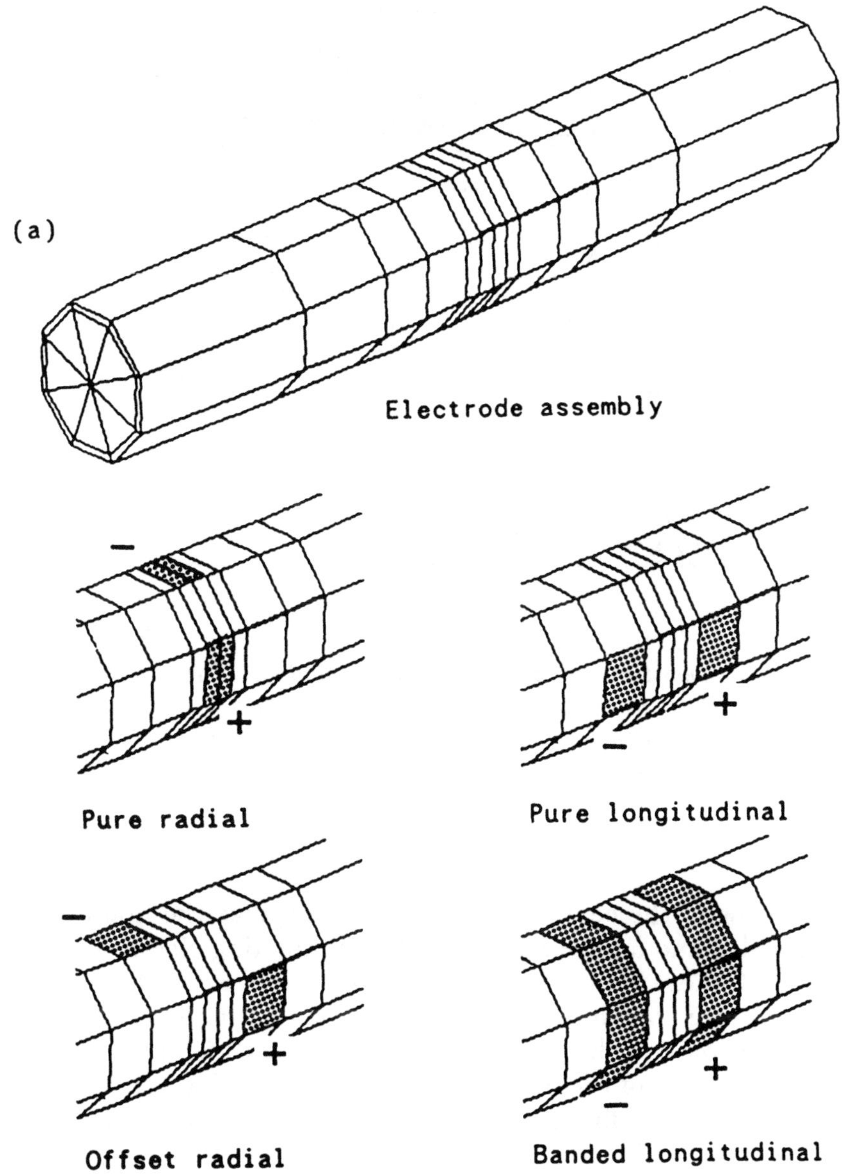

Fig. 12.27 *Bipolar electrodes. (a) 12 layers of fininte elements are used to describe the electrode. Pure radial electrodes have the sharpest focus effect, followed by the offset radial, the pure longitudinal, and the banded longitudinal electrodes.*

(From Finley et al., 1990)

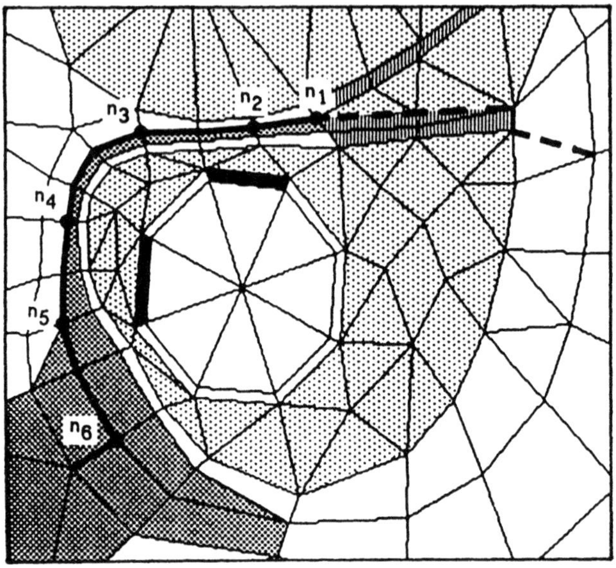

Fig. 12.28 Detailed diagram of the cross section of Fig. 12.26, including the position of auditory nerve nodes. The nerve starts at node n_1 after passing the habenula perforata. Note that the distances between the nodes differ from the geometry shown in Fig. 12.25. A bipolar, pure radial electrode is assumed to be implanted in the scala tympani.

(From Finley et al., 1990)

detailed view of the central part, we can recognize the main parts of the cochlea and a pure radial bipolar stimulating electrode (Fig. 12.26(b)). Other types of bipolar electrodes used for multi channel stimulation are shown in Fig. 12.27. With this model FINLEY *et al.* (1990) have calculated the influence of the different electrode types at the nodes of the primary auditory nerve. Fig. 12.28 shows a central section of finite elements, which were used to calculate the potential distribution, produced by different types of bipolar electrodes. Potential distribution and the value of the activating function in the nodes of Ranvier help in analyzing the influence of electrode geometry on nerve reactions.

The upper part of Fig. 12.29 shows the isopotentials produced by a pure radial dipole electrode, located similarly to Fig. 12.28. The influence of a pulsatile signal on 7 fibers, marked by f_1, f_2, ... f_7 at different distances of the basilar membrane, is investigated. Since the situation is symmetrical, there are only 4

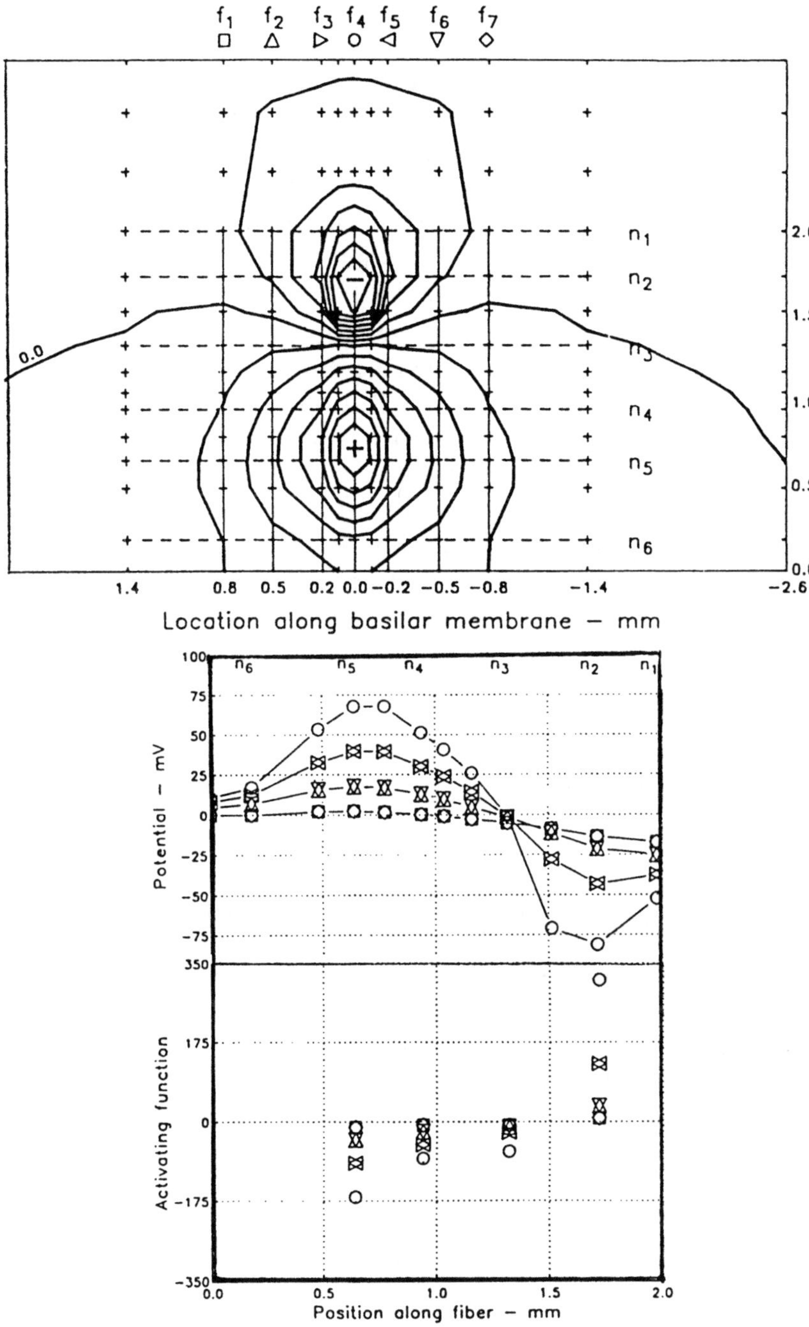

Fig. 12.29 *Potential distribution and activating function for pure radial electrodes.*

(From Finley et al., 1990)

different reactions. Every fiber is represented here by 6 nodes.

For simplicity it is assumed that all the nodes lie in a plane which contains the cochlea axis. This is only an approach of the real situation. Furthermore, in profoundly deaf people extensive loss of afferent auditory fibers often occurs. However, even when the fibers are present, the peripheral processes of the auditory nerves can be absent; therefor, target element for electrical stimulation becomes only the centrally projecting axon.

The electrode has the strongest influence on the central fiber f_4. By inspection of the activating function it turns out that a pulsatile signal can generate excitation only at node 2, but detailed computation is necessary to predict the overall behavior of the axon. It is possible that the following hyperpolarized part will block this excitation. But usually, after a pulsatile signal, another one with the opposite polarization will be sent in order to avoid charge accumulation. This signal probably evokes an AP.

Since the high value of the activating function correlates with a low threshold, and since the n_2 value of the activating function diminishes rapidly with increasing distances between fiber and electrode, very high thresholds would be expected at all locations within the cochlea except near the radial electrodes (FINLEY *et al.*, 1990). This single unit threshold behavior of radial bipolar electrodes has been observed by VAN DEN HONERT & STYPULKOWSKY (1987a) in cat experiments.

We find quite a different situation for stimulation with pure longitudinal electrodes (Fig. 12.30). The strongest stimulating reactions are expected in fibers f_2 and f_3, but the fibers closest to the anodic electrode (f_5, f_6) are only moderately stimulated at the most peripheral nodes n_2. For fibers nearest to the electrode pair, two threshold dips would be expected. The one nearest the basal cathode (f_2, f_3) would be relatively deep, but not as deep as in the case of pure radial electrodes. Another, even shallower dip, is expected near the apical electrode (f_5, f_6).

Unfortunately, detailed computations on fiber reactions have not been achieved yet. Therefore, the interested reader will find an extensive field for research on this subject. It should be noted that, in this case, the activating function (and also the master equation 7.5 for calculating the nodal reaction) should be used in a more generalized form, because fiber diameters and internodal lengths are not constant in the first segments of the fiber.

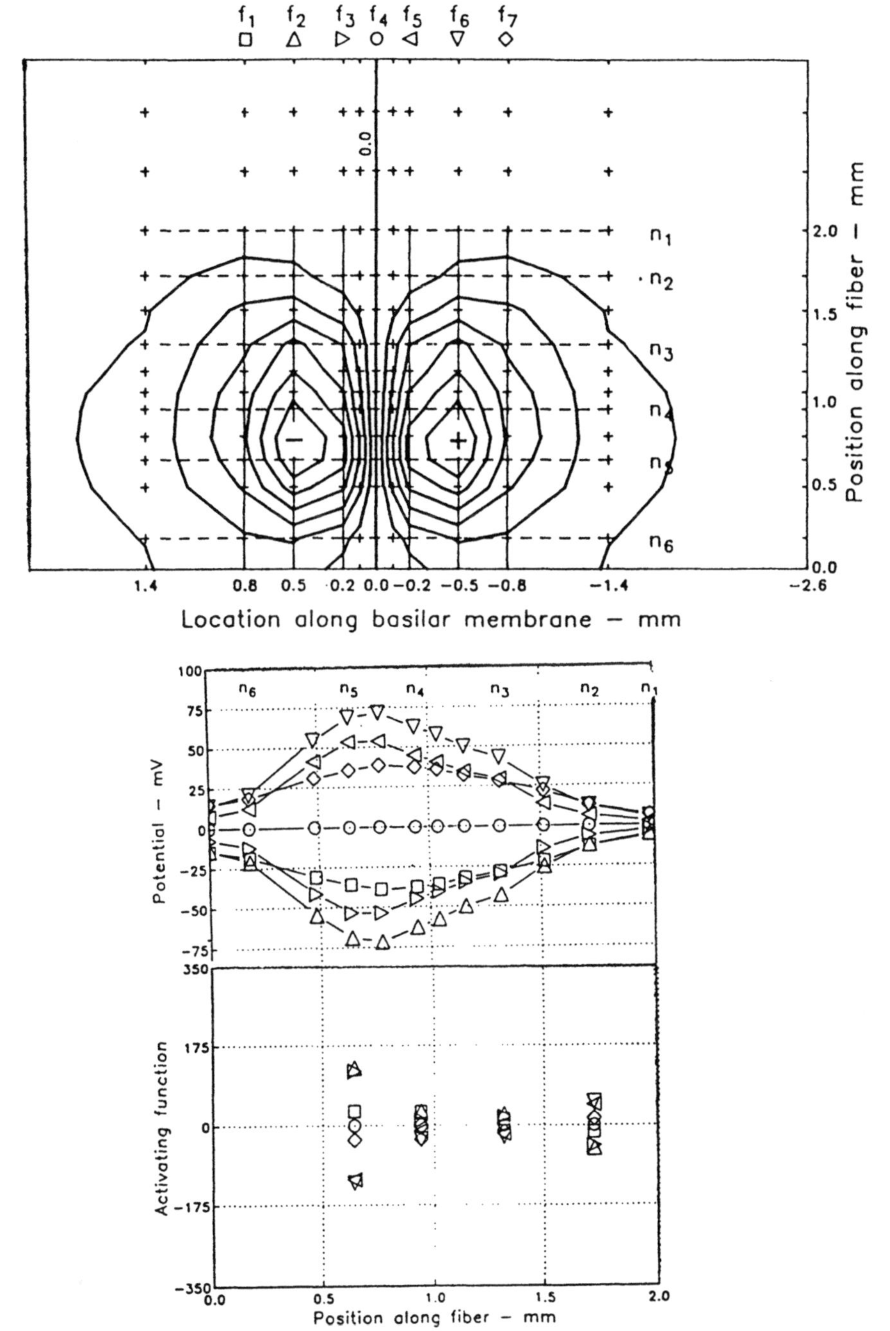

Fig. 12.30 *Potential distribution and activating function for pure longitudinal electrodes.*

(From Finley et al., 1990)

The primary auditory nerve has a special structure with relatively long internodal distances. These long myelinated segments should not be neglected in a more adequate model. In order to find a set of equations, we will use an electric network, which simulates the internodal segments by a separate circuit (Fig. 12.31). Segment 0 simulates the unmyelinated part at the peripheral end of the axon (Fig. 12.25), the following even segments symbolize the nodes of RANVIER, and the odd segments symbolize the myelinated internodes plus the cell body.

From the electric network we can derive the equations as described in Chapter 6 and 7. With the same symbols used in eqn. 6.1, we obtain (without injected stimulus current) for a node:

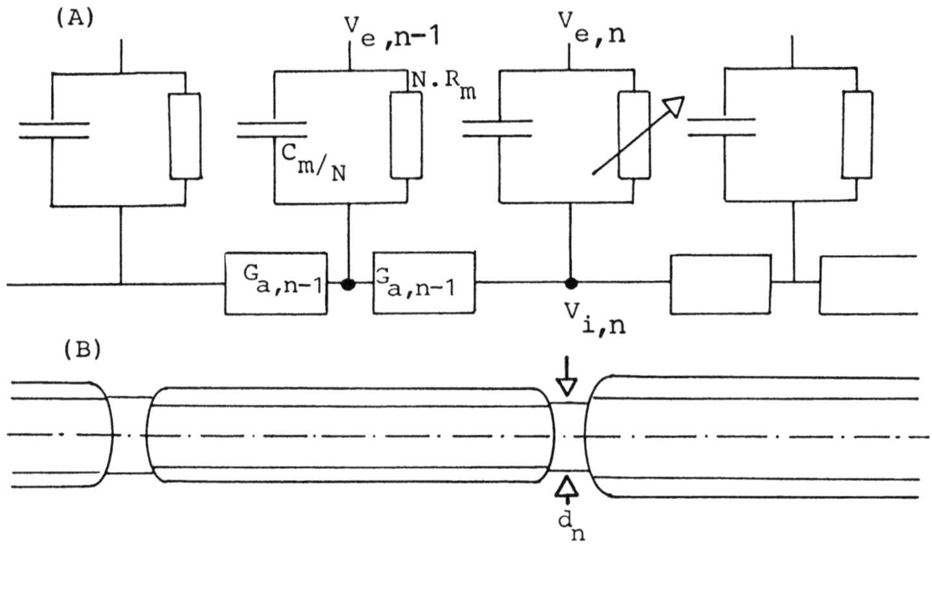

Fig. 12.31 Electric network for fibers with changing diameters. The network (A) consists of active circuits with nonlinear membrane conductances. These define the ionic current at the nodes and the passive circuits with constant resistance resulting from N layers of Schwann cell membranes. Every passive element symbolizes one internodal segment. The voltages V_e and V_i are average values of the segments. Passive segment length: Δx, active segment length: L; diameter d varies with segments.

$$C_m \frac{d(V_{i,n} - V_{e,n})}{dt} + I_{i,n} + G_{a,n-1}(V_{i,n} - V_{i,n-1}) + G_{a,n+1}(V_{i,n} - V_{i,n+1}) = 0 \tag{12.1}$$

Axonal conductance G_a, external and internal potential V_e and V_i as well as fiber diameter depend now on element number n.

In order to use the current densities of the additional membrane models we introduce

$$C_m = c_m \pi d_n L \tag{12.2}$$

$$G_{a,n-1} = \frac{\pi d_{n-1}^2}{4\rho_i \frac{\Delta x_{n-1}}{2}} \tag{12.3}$$

plus a similar expression for $G_{a,n+1}$. Instead of (6.6) we obtain the following equation for even n which marks a nodal section:

$$\dot{V}_n = \left\{ -i_{ionic,n} + \frac{1}{2Ld_n\rho_i} \left[\frac{d_{n-1}^2}{\Delta x_{n-1}} \cdot V_{n-1} - \left(\frac{d_{n-1}^2}{\Delta x_{n-1}} + \frac{d_{n+1}^2}{\Delta x_{n+1}} \right) \cdot V_n + \frac{d_{n+1}^2}{\Delta x_{n+1}} \cdot V_{n+1} + \frac{d_{n-1}^2}{\Delta x_{n-1}} \cdot V_{e,n-1} - \left(\frac{d_{n-1}^2}{\Delta x_{n-1}} + \frac{d_{n+1}^2}{\Delta x_{n+1}} \right) \cdot V_{e,n} + \frac{d_{n+1}^2}{\Delta x_{n+1}} \cdot V_{e,n+1} \right] \right\} / c_m \tag{12.4}$$

Note that there are three different diameters involved: d_n is the nodal diameter, d_{n-1} and d_{n+1} are the diameters of the left and right internodal element, respectively. Asssuming $d_0 = d_1$ for the starting element (n=0), eqn. 12.4 reduces to

$$\dot{V}_0 = \left\{ -i_{ionic,0} + \frac{d_1}{2L_0\rho_i\Delta x_1} [V_1 - V_0 + V_{e,1} - V_{e,0}] \right\} / c_m \tag{12.5}$$

The internodal voltages are calculated for an odd n by

$$\dot{V}_n = \left\{ -\frac{V_n}{N \cdot r_m} + \frac{d_n}{2\rho_i \Delta x_n^2} [V_{n-1} - 2V_n + V_{n+1} + V_{e,n-1} - 2V_{e,n} + V_{e,n+1}] \right\} / \frac{c_m}{N} \tag{12.6}$$

where r_m denotes membrane resistance in $k\Omega$ per cm^2 and N is the number of membranes wrapped around the myelinated part of the axon.

REFERENCES

Abbas P.J. 1988. Electrophysiology of the auditory system. Clin. Phys. Physiol. Meas. 9, 1-31

Adelman W.J. and Gilbert D.L. 1964. Internally perfused squid axons studied under voltage clamp conditions. I. Meth. J. Cell. Comp. Physiol. 64, 423-428

Adrian E.D. 1914. The all or none principle in nerve. J. Physiol. 47, 460-474

Aidley D.J. 1978. *The physiolgy of excitable cells.* Cambridge: University Press

Allen J.B. 1985. Cochlear modeling. IEEE ASSP Magazine. Jan. 1985, 3-29

Altman K.W. and Plonsey R. 1990. Point source nerve bundle stimulation: effects of fiber diameter and depth on simulated excitation. IEEE Trans. Biomed. Eng. BME-37, 688-698

Altschuster R.A., Bobbin, R.B. and Hoffman D.W. [Ed.] 1986. *Neurobiology of Hearing: The Cochlea.* Raven Press. New York

Arbuthnott E.R., Boyd I.A. and Kalu K.U. 1980 Ultrastructural dimensions of myelinated peripheral nerve fibres in the cat and their relation to conduction velocity. J. Physiol. 308, 125-127

Arrhenius Sv. 1887. Über die Dissociation in Wasser gelöster Stoffe. Zeitschr. physik. Chemie 1, 631-648

Bak M., Girvin J.P., Hambrecht F.T., Kufta C.V., Loeb G.E. and Schmidt E.M. 1990. Visual sensations produced by intracortical microstimulation on the human occipital cortex. Med. & Biol. Eng. & Comput. 257-259

Baker P.F., Hodgkin A.L. and Shaw T.I. 1961. Replacement of the protoplasm of a giant nerve fibre with artificial solutions. Nature 190, 885-887

Bartlett J.R. and Doty R.W. 1980. An exploration of the ability of macaques to detect microstimulation of striate cortex. Acta Neurobiol. Exp. 40, 713-727

Beard G.M. and Rockwell A.D. 1878. *On the medical and surgical uses of electricity.* William Wood & Co. New York

Beeler G.W. and Reuter H. 1974. Reconstruction of the action potential of ventricular myocardial fibers. J. Physiol. 268, 177-210

Bèkèsy G. 1960. *Experiments in Hearing*, McGraw-Hill, New York

BeMent S.L. and Ranck J.B. 1969a. A quantitative study of electrical stimulation of central myelinated fibers. Exp. Neurol. 24, 147-170

BeMent S.L. and Ranck J.B. 1969b. A model for electrical stimulation of central myelinated fibers with monopolar electrodes. Exp. Neurol. 24, 171-186

Bernstein J. 1868. Über den zeitlichen Verlauf der negativen Schwankungen des Nervenstroms. Arch. Ges. Physiol. 1, 173-207

Bernstein J. 1902. Untersuchungen zur Thermodynamik der biologischen Ströme. Pflügers Arch. 92, 521-562

Bernstein J. 1912. *Elektrobiologie.* Vieweg, Braunschweig (Germany)

Bezanilla F. 1987. Single sodium channels from the squid axon. Biophys. J. 52, 1087-1090

Black R.C. and Clark G.M. 1980. Differential electrical excitation of the auditory nerve. J. Acoust. Soc. Am. 67, 868-874

Blair E.A. and Erlanger J. 1933. A comparison of the characteristics of axons through their individual electrical responses. Am. J. Physiol. 106, 524-564

Bogoslovskaya L.S., Lyubinskii I.A., Pozin N.V., Putsillo Y.V., Shmelev L.A. and Shura-Bura T.M. 1973. Spread of excitation along a fiber with local inhomogeneities (results of modelling). Biophysics 18, 944-948

Bonhoeffer K.F. 1943. Zur Theorie des elektrischen Reizes. Naturwissenschaften 31, 270-275

Bonhoeffer K.F. 1948. Activation of passive iron as a model for the excitation of nerve. J. Gen. Physiol. 32, 69-91

Bonhoeffer K.F. 1953. Modelle der Nervenerregung. Naturwissenschaften 40, 301-311

Bostock H. and Sears T.A. 1978. The internodal axon membrane: Electrical excitability and continuous conduction in segmental demyelination. J. Physiol. 280, 273-301

Bostock H., Sears T.A. and Sherratt R.M. 1981. The effects of 4-aminopyridine and tetraethylammonium ions on normal and demyelinated mammalian nerve fibres. J. Physiol. 313, 301-315

Bretscher M.S. 1985. The molecules of the cell membrane. Scientific Am. 253/4 86-90

Bretscher M.S. and Raff M.C. 1975. Mammalian plasma membranes. Nature 258, 43-49

Brindley G.S. and Lewin W. 1968. The sensations produced by electrical stimulation of the visual cortex. J. Physiol. 196, 479-493

Brugge J.F., Anderson D.J., Hind J.E. and Rose J.E. 1969. Time structure of discharges in single auditory nerve fibers of the squirrel monkey in response to complex periodic sounds. J. Neurophysiol. 32, 386-401

Brummer S.B., Robblee L.S and Hambrecht F.T. 1983. Criteria for selecting electrodes for electrical stimulation: theoretical and practical considerations. In cochlear prostheses. Parkins C.W. and Anderson S.W. (Ed.) Ann. New York academy of sciences. Vol. 405, 159-171

Bütikofer R. and Lawrence P.D. 1978. Electrocuteneous nerve stimulation – 1: Model and experiment. IEEE Trans. Biomed. Eng. BME-25, 526-531

Carpenter R.H.S. 1984. *Neurophysiology.* Arnold, London

Chiu S.Y., Ritchie J.M., Rogart R.B. and Stagg D. 1979. A quantitative description of membrane currents in rabbit myelinated nerve. J. Physiol. 292, 149-166

Chiu S.Y. and Ritchie J.M. 1981. Evidence for the presence of potassium channels in the paranodal region of acutely demyelinated mammalian nerve fibers. J. Physiol. 313, 415-437

Clark J. and Plonsey R. 1968. The extracellular potential field of the single active nerve fiber in a volumne conductor. Biophys. J. 8, 842-864

Coburn B. 1985. A theoretical study of epidural stimulation of the spinal cord. II. Efects on long myelinated fibers. IEEE-Trans. Biomed. Eng. BME-32, 978-986

Coburn B. 1989. Neural modeling in electrical stimulation. Critical Reviews Biomed. Eng. 17, 133-178

Cohen H. 1976. Mathematical developments in Hodgkin-Huxley theory and its applications. Lectures on Math. in the Life Sciences 8, 89-124

Cole K.S. 1968. *Membranes, ions and impulses.* University of California Press, Berkeley

Cole K.S. and Curtis H.J. 1939. Electric impedance of the squid giant axon during activity. J. Gen. Physiol. 22, 649-670

Cole K.S. and Hodgkin A.L. 1939. Membrane and protoplasm resistance in the squid giant axon J. Gen. Physiol. 22, 671-687

Colombo J. and Parkins C.W., 1987. A model of electrical excitation of the mammalian auditory-nerve neuron. Hearing Res. 31, 287-312

Connor J.A. and Stevens C.F. 1971a. Inward and delayed outward membrane currents in isolated neural somata under voltage clamp. J. Physiol. 213, 1-19

Connor J.A. and Stevens C.F. 1971b. Voltage clamp studies of a transient outward membrane current in gastropod neutral somata. J. Physiol. 213, 21-30

Connor J.A. and Stevens C.F. 1971c. Predictions of repetitive firing behaviour from voltage clamp data on an isolated neurone soma. J. Physiol. 213, 31-53

Connor J.A., Walter D. and McKnown R. 1977. Neural repetitive firing. Modification of the Hodgkin-Huxley axon suggested by experimental results from crustacean axons. Biophys. J. 18, 81-102

Conti F., De Felice L.J. and Wanke E. 1975. Potassium and sodium ion current noise in the membrane of the squid giant axon. J. Physiol. 248, 45-82

Cooley J.W. and Dodge F.A. 1966. Digital computer solutions for excitation and propagation of the nerve impulse. Biophys. J. 6, 583-599

Crago P.E., Chizeck H.J., Neuman R. and Hambrecht F.T. 1986. Sensors for use with functional neuromuscular stimulation. IEEE Trans. Biomed. Eng. BME-33, 256-268

Crawford A.C. and Fettiplace R. 1985. The mechanical properties of ciliary bundles of turtle cochlear hair cells. J. Physiol. 364, 359-379

Cronin J. 1981. *Mathematics of cell electrophysiolgy.* Dekker, New York

Dallos P. 1973. *The Auditory Periphery.* Academic, New York

Davis H. 1970. Anatomy and physiology of the auditory system. (In *Hearing and Deafness*, Ed. Davis H. and Silverman S.R., 3rd ed., Holt, Rinehart and Winston. 47-82

Djourno A. and Eyries C. 1957. Prosthèse auditive par eratation électrique à distance du nerf sensoriel à l' aide d'un bobinage inclus á demeure. Presse Med. 65, 1917

Dobelle W.H. and Mladejovsky M.G. 1974. Phosphenes produced by electrical stimulation of human occipital cortex and their application to the development of a prosthesis for the blind. J. Physiol. 243, 553-576

Dobelle W.H., Mladejovsky M.G., Evans J.K., Roberts T. and Girvin J.P. 1976. 'Braille' reading by a blind volunteer by visual cortex stimulation. Nature 259, 111-112

Dobie R.A. and Dillier N. 1985. Some aspects of temporal coding for single-channel electrical stimulation of the cochlea. Hearing Research 18, 41-55

Dodge F.A. and Frankenhaeuser B. 1958. Membrane currents in isolated frog nerve fibre und voltage clamp conditions. J. Physiol. 143, 76-90

Durand D., Ferguson A.S. and Dalbasti T. 1989. Induced electric fields by magnetic stimulation in non-homogeneous conducting media. Proc. 11th Ann. Conf. IEEE-EMBS 1252-1253

Ettinger K.V., Ahamed Z.M. and Brondo J. 1987. Microelectronic implants for neural bidirectional interfacing. IEEE 9-th ann. conf. Eng. Med. Biol. Soc. Boston. 488-489

Fang Z.P. and Mortimer J.T. 1987. A method for attaining natural recruitment order in artificially activated muscles. IEEE 9-th ann. conf. Eng. Med. Biol. Soc. Boston. 657-658

Ferguson A.S., Durand D. and Dalbasti T. 1989 Optimization of coil design for neuronal excitation by magnetic stimulation. Proc. 11th Ann. Conf. IEEE-EMBS 1254-1255

Ferguson A.S., Sweeney J.D., Durand D. and Mortimer J.T. 1987. Finite difference modeling of nerve cuff electric fields. Proc. 9th Ann. Conf. IEEE-EMBS 1579-1580

Finley C.C., Wilson B.S. and White M.W. 1990. Models of neural responsiveness to electrical stimulation. In Cochlear implants: models of the electrically stimulated ear (Ed. Miller J.M. and Spelman F.A.) Springer, New York. 55-96

Fitzhugh R. 1961. Impulses and physiological states in theoretical models of nerve membrane. Biophys. J. 1, 445-466

Fitzhugh R. 1969. Mathematical models of excitation and propagation in nerve. In biological engineering (Ed. H.P. Schwan) McGraw-Hill, New York. 1-85

Frankenhaeuser B. and Hodgkin A.L. 1956. The after-effects of impulses in the giant fibres of *Loligo*. J. Physiol. 131, 341-346

Frankenhaeuser B. and Huxley A.L. 1964. The action potential in the myelinated nerve fibre of *Xenopus Laevis* as computed on the basis of voltage clamp data. J. Physiol. 171, 302-315

Frankenhaeuser B. and Moore L.E. 1963. The effect of temperature on the sodium and potassium permeability changes in myelinated nerve fibers of *Xenopus Laevis*. J. Physiol. 169, 431-437

Freed W.J., DeMedinaceli L., et al. 1985 (March). Promoting functional plasticity in the demaged nervous system. Science, 227, 1544-1552

Fricke H. 1923. The electric capacity of cell suspension. Physic. Rev. 21, 708-709

Fritsch G.T. 1887. *Die elektrischen Fische.* Von Veit, Leipzig

Galambos R. and Davis H. 1943. The response of single auditory nerve fibers to acoustic stimulation. J. Neurophysiol. 6, 39-57

Ganapathy N. and Clark J.W. jr. 1987. Extracellular currents and potentials of the active myelianted nerve fiber. Biophys. J. 52, 749-761

Galvani L. 1791. De viribus electricitatis in motu musculari. Commentarius de bononiensi scientiarum et artium instituto atque academia commentarii 7, 363

Geddes L.A. and Baker L.E. 1967. The specific resistance of biological material - A compendium of data for the biomedical engineer and physiologist. Med. Biol. Eng. 5, 271-293

Gelfand S.A. 1981. *Hearing* M. Dekker, New York Glass I. 1985. Responses of cochlear nucleus units to electrical stimulation through a cochlear prosthesis: Channel interaction. Hearing Res. 17, 115-126

Goldman D.E. 1943. Potential, impedance and rectification in membranes. J. Gen. Physiol. 27, 37-60

Gorman A.L.F. and Thomas M.V. 1978. Changes in the intracellular concentrations of free calcium ions in a pacemaker neuron measured with the metallochromic indicator dye Arsenazo III. J. Physiol. 275, 357-376

Guttman R. and Barnhill R. 1970. Oscillations and repetitive firing in squid axons. Comparison of experiments with computations. J. gen.

Physiol. 55, 104-118

Hamil O.P. and Sakmann B. 1981. Multiple conductance states of single acetylcholine receptor channels in embryonic muscle cells. Nature 294, 462-464

Hartmann R. and Klinke R. 1990. Response characteristics of nerve fibers to patterned electrical stimulation. In *Cochlear implants.* (Ed. Miller J.M. and Spelman F.A.) Springer. New York. 135-160

Hartmann R., Topp G. and Klinke R. 1984. Discharge patterns of cat primary auditory fibers with electrical stimulation of the cochlea. Hearing Res. 13, 47-62

Helmholtz H. *Die Lehre von den Tonempfindungen.* (Engl. translation: A.J. Ellis, On the sensations of tones, 1885) Henneman E., Somjen G. and Carpenter D.O. 1965a. Functional significance of cell size in spinal motoneurons. J. Neurophysiol. 28, 560-580

Henneman E., Somjen G. and Carpenter D.O. 1965b. Excitability and inhibitibility of motoneurons of different sizes. J. Neurophysiol. 28, 599-619

Henneman E. 1981. Recruitment of motoneurons: the size principle. In *Motor unit types, recruitment and plasticity in health and disease.* (Ed. J.E. Desmedt, S. Karger) Prog. clin. Neurophysiol. 9.

Hermann L. 1905. Beiträge zur Physiologie und Physik des Nerven. Arch. Ges. Physiol. 109, 95-144

Hille B. 1971. The permeability of the sodium channel to organic cations in myelinated nerve. J. gen. Physiol. 58, 599-619

Hochmair-Desoyer I.J., Hochmair E.S., Burian K. and Fischer R.E. 1981. Four years of experience with cochlear prostheses. Med. Prog. Technol. 8, 107-119

Hochmair-Desoyer I.J., Hochmair E.S., Motz H. and Rattay F. 1984. A model for the electrostimulation of the nervus acusticus. Neuroscience 13, 553-562

Hodgkin A.L. 1964. *The conduction of the nervous impulse.* Springfield, Ill. Thomas

Hodgkin A.L. and Huxley A.F. 1939. Action potentials recorded from inside a nerve fibre. Nature 144, 710-711

Hodgkin A.L. and Huxley A.F. 1945. Resting and action potentials in single nerve fibres. J. Physiol. 104, 176-195

Hodgkin A.L. and Huxley A.F. 1952a. The components of membrane conductance in giant axon of *Loligo*. J. Physiol. 116, 473-496

Hodgkin A.L. and Huxley A.F. 1952b. A quantitative description of membrane current and its application to conduction and excitation in nerve. J. Physiol. 117, 500-544

Hodgkin A.L. and Katz B. 1949. The effect of temperature on the electrical activity of the giant axon of the squid. J. Physiol. 109, 240-249

Hodgkin A.L. and Rushton W.A.H. 1946. The electric constants of a crustacean nerve fibre. Proc. Roy. Soc. London B 133, 444-479

Holden A.V. 1980. The mathematics of excitation. In Biomathematics 1980. (Ed. L.M. Ricciardi and A.C. Scott) Noth Holland

Holle J., Frey M., Gruber H., Kern H., Stöhr H. and Thoma H. 1984. Functional electrostimulation of paraplegics: Experimental investigations and first clinical experience with an implantable stimulation device. Orthopedics 7, 1145-1155

Holle J., Moritz E., Thoma H. and Lischka A. 1974. Die Karusselstimulation, eine neue Methode zur elektrophrenischen Langzeitbeatmung. Wiener klinische Wochenschrift 86, 23-27

van den Honert C. and Mortimer J.T. 1979. Generation of unidirectionally propagated action potentials in peripheral nerve by brief stimuli. Science 206, 1311-1312

van den Honert C. and Mortimer J.T. 1981. A technique for collision block of peripheral nerve: Single stimulus analysis. IEEE Trans. Biomed. Eng. BME 28, 373-378

van den Honert C. and Mortimer J.T. 1981. A technique for collision block of peripheral nerve: Frequency dependence. IEEE Trans. Biomed. Eng. BME 28, 379-382

Hoorweg J.L. 1892. Condensatorentladung und Auseinandersetzung mit du Bois Reymond. Pflügers Arch. 52, 87-108

Horáckova M., Nonner W. and Stämpfli R. 1968. Action potentials and voltage clamp currents of single rat Ranvier nodes. Pro. Int. Union Physiol. Sci. 7, 198

House W.F., Berliner K.I., Crary W.G. et al. 1976. Cochlear implants. Ann. Otol. Rhinol. Laryngol. 85 (Suppl. 27), 1-93

Hubel D.H. 1982. Exploration of the primary visual cortex. Nature 299, 515-524

Hubel D.H. and Wiesel T.N. 1979. Brain mechanisms of vision. Sci. Am. 241, 150-162

Hudspeth A.J. and Corey D.P. 1977. Sensitivity, polarity, and conductance change in the response of vertebrate hair cells to controlled mechanical stimuli. Proc. Natl. Acad. Sci. USA 74, 2407-2411

Hudspeth A.J. 1989. How the ear's works work. Nature 341, 397-404

Huneus-Cox F., Fernandez H.L. and Smith B.H. 1966. Effects of redox and sulfhydryl reagents on the bioelectric properties of the giant axon of the squid. Biophys. J. 6, 675-689

Huxley A.F. 1959. Ion movements during nerve activity. Ann. N.Y. Acad. Sci. 81, 221-246

Huxley A.F. and Stämpfli R. 1949. Evidence for saltatory conduction in peripheral myelinated nerve fibres. J. Physiol. 108, 315-339

Jack J.J.B., Noble D. and Tsien R.W. 1983. *Electric current flow in excitable cells.* Clarendon Press, Oxford

Jackson J.D. 1962. *Classical electrodynamics.* Wiley, New York

Jankowska E. and Roberts W.J. 1972. An electrophysiological demonstration of the axonal projections of single interneurons in the cat. J. Physiol. 222, 597-622

Jankowska E. Smith D.O. 1973. Antidromic activation of Renshaw cells and the distribution of their axonal branches. Acta phys. scand. 88, 198-214

Javel E. 1986. Basic response properties of auditory nerve fiber. In *Neurobiology of Hearing: The Cochlea.* (Ed.: Altschuster R.A., Bobbin, R.B. and Hoffman D.W.) Raven Press. New York

Katz B. 1966. *Nerve, muscle and synapse.* McGraw-Hill, New York

Katz B. and Miledi R. 1972. The statistical nature of the acetylcholine potential and its molucular components. J. Physiol. 224, 665-699

Kelvin Lord (William Thompson) 1855. On the theory of the electric telegraph. Proc. Roy. Soc. London B. 135, 506-534

Kernell D. 1986. Organization and properties of spinal motoneurons and motor units. Progress in Brain Research 64, 21-30

Keynes R.D. and Aidley D.J. 1983. *Nerve and muscle.* Cambridge: University Press

Khanna, S.M. and Leonard D.G.B. 1982. Basilar membrane tuning in the cat. Science 215, 305-306

Khodorov B.I. 1974 *The problem of excitability.* Plenum, New York

Khodorov B.I., Timin Y.N., Vilenkin S.Y. and Gol'ko F.B. 1969. Theoretical analysis of the mechanisms of conduction of a nerve pulse over an inhomogeneous axon. I. conduction through a portion with increased diameter. Biophysics 14, 323-335

Khodorov B.I., Timin Y.N., Pozin N.V. and Shmelev L.A. 1971. Theoretical analysis of the mechanisms of conduction of a nerve pulse over an inhomogeneous axon. IV. conduction of a series of impulses through a portion of the fibre with increased diameter. Biophysics 16, 96-104

Kiang N.Y.S., Liberman M.C. and Levin R.A. 1976. Auditory-nerve activity in cats exposed to ototoxic drugs and high-intensity sounds. Ann. Otol. Rhinol. Laryngol. 85, 752-768

Kiang N.Y.S., Watanable T., Thomas E.C. and Clark L.F. 1965. *Discharge patterns of single fibers in the cat's auditory nerve.* MIT Research monograph No. 35. MIT Press, Cambridge, MA

Klumpp R.G. and Eady H.R. 1956. Some measurements of interaural time difference thresholds. J. Acoust. Soc. Am. 28, 859-860

Ko W.H., Liang S.P. and Fung C.D.F. 1977. Design of radio-frequency powered coils for implant instruments. Med. Bio. Eng. Comput. 15, 634-640

Kohlrausch F. 1876. Über das Leitungsvermögen der in Wasser gelösten Elektrolyte im Zusammenhang mit der Wanderung ihrer Bestandtheile. Nachr. K. Ges. der Wiss. (Göttingen) 213-244

Kokis J.D. and Waxman S.G. 1987. Ionic channel organization of normal and regenerating mammalian axons. In *Progress in Brain Research* Ed. Seil F.J., Herbert E. and Carlson. Amsterdam Elsevier 71, 89-101

Konishi T. and Nielsen D.W. 1978. Temporal relationship between basilar membrane motion and nerve impulse initiation in auditory nerve fibers of guinea pigs. Jpn. J. Physiol. 28, 291-307

Kovacs G.T.A., Stepheanides M.C., Knapp W.R., McVittie J.P., Rosen J.M. 1987. Design of two-dimensional neural prosthesis microelectrode arrays. IEEE 9-th ann. conf. Eng. Med. Biol. Soc. Boston. 1034-1035

Kralj A., Bajd T., Turk R. and Benko H. 1986. Posture switching for prolonging functional electrical stimulation standing in paraplegic patients. Paraplegia 24, 221-230

Kralj A., Bajd T., Turk R., Krajnik J. and Benko H. 1983. Gait restoration in paraplegic patient: a feasibility demonstration using multichannel surface electrode FES. J. of Rehabilitation R & D, 20, 3-19

Kuffler S.W., Nicholls J.G. and Martin A.R. 1984. *From neuron to brain.* 2nd edn. Sinauer, Sunderland, Massachusetts

Lapicque L. 1907. Recheres quantitatifs sur l'excitation electrique des nerfs traitee comme une polarisation. J. Physiol. Paris 9, 622-635

Lim D.J. 1986. Functional structure of the organ od Corti: a review. Hearing Res. 22, 117-146

Liberman M.C. 1978. Auditory-nerve activity from cats raised in a low-noise chamber. J. Acoust. Soc. Am. 63, 442-455

Loeb G.E., McHardy J., Kelliher E.M. and Brummer S.B. 1982. Neural prostheses. Biocompatibility in clinical practice, vol. II. Ed. D.F. Williams, CRC press, Boca Raton, Florida, 124-149

Loeb G.E., White M.W. and Jenkins W.M. 1983. Biophysical considerations in electrical stimulation of the auditory nervous system. In cochlear prostheses. Ed. C.W. Parkins and S.W. Anderson. Ann. N.Y. Acad. Sci. 405, 123-136

Lucas K. 1908/1909. The 'all or none' contraction of the amphibian skeletal muscle fibre. J. Physiol. 38, 113-133

McAllister R.E., Noble D. and Tsien R.W. 1975. Reconstruction of the electrical activity of cardiac Purkinje fibres. J. Physiol. 251, 1-59

McNeal D.R. 1976. Analysis of a model for excitation of myelinated nerve. IEEE Trans. Biomed. Eng. BME-23, 329-337

McNeal D.R. 1977. 2000 years of electrical stimulation. In Functional electrical stimulation - application in neural prostheses. Ed. F.T. Hambrecht and J.B. Reswick, Dekkeer Inc. New York. 3-33

Meadows P.M., McNeal D.R., Su N.Y. and Tu W.W. 1987. Development of an implantable electrical stimulation system for gait applications in stroke and spinal cord injured patients. IEEE 9-th ann. conf. Eng. Med. Biol. Soc. 618-619

Meves H. 1984. Hodgkin-Huxley: thirty years after. Current Topics in Membrane and Transport 22, 279-331

Michelson R.P. 1971. Electrical stimulation of the human cochlea. Arch. Otolaryngol. 93, 317-323

Miller J.M. and Spelman F.A. (Ed.) 1990. *Cochlear implants: models of the electrically stimulated ear.* Springer, New York.

Møller A.R. (Ed.) 1973. *Basic Mechanisms in Hearing.* Academic Press, New York

Møller A.R. 1983. *Auditory Physiology.* Academic Press, New York

Moritz E., Holle J., Thoma H. and Stöhr H. 1983. Further experiences with electrophrenic respiration. Proc. 1st Vienna int. workshop on functional electrostimulation 6.5

Mortimer J.T. 1981. Motor prostheses. Handbook of physiolgy: The nervous system, vol II. Ed. V.B. Brooks, American Physiological Society, Bethesda, Maryland, 155-187

Mortimer J.T., Sweeney J.D., Bodner D.R. and Ferguson A.S. 1988 An implantable cuff electrode for collision block of pudendal nerve motor activity. IEEE 10-th ann. conf. Eng. Med. Biol. Soc. 1523-1524

Motz H. und Rattay F. 1986. A study of the application of the Hodgkin-Huxley and the Frankenhaeuser-Huxley model for electrostimulation of the acoustic nerve. Neuroscience 18, 699-712

Motz H. and Rattay F. 1987. Signal processing for electrostimulated ear prostheses based on simulated nerve response. Perception 16, 777-784

Moxon E.C. 1971. Neural and mechanical responses to electric stimulation of the cat's inner ear. Thesis. MIT. Cambridge, Mass.

Mullar A. and Hunsperger B. 1967. Reversible Blockierung der Erregungsleitung in Nerven durch Mittelfrequenz Dauerstrom. Helv. Physiol. Acta 25, 211-213

Nathan R.H. 1986. Programmed gripping modes by electrical stimulation of the upper limb. Proc. 2nd Vienna int. workshop on functional electrostimulation. 109-116

Nathan R.H. 1990. FNS of the upper limb: targeting the forearm muscles for surface stimulation. Med. & Biol. Eng. & Comput. 249-256

Naples G.G., Mortimer J.T., Scheiner A. and Sweeney J.D. 1988. A spiral nerve cuff electrode for peripheral nerve stimulation. IEEE-Trans. Biomed. Eng. BME-35, 905-916

Neher E. and Sakmann B. 1976. Single-channel currents recorded from denervated frog muscle fibres. Nature 260, 799-802

Nernst W. 1888. Zur Kinetik der in Lösung befindlichen Körper: Theorie der Diffusion. Zeitschr. physik. Chemie 2, 613-637

Nernst W. 1889. Die elektromotorische Wirksamkeit der Ionen. Zeitschr. physik. Chem. 4, 129-181

Nernst W. 1908. Zur Theorie der elektrischen Reizung. Arch. Ges. Physiol. 122, 275-314

Noble D. 1979. *The initiation of the heartbeat.* 2nd edn. Oxford university press.

Oikawa T., Spyropoulos C.S., Tasaki I. and Teorell T. 1961. Methods for perfusing the giant axon of *Loligo pealii.* Acta physiol. Scand. 52, 195-196

Parkins C.W. and Anderson S.W. (Ed.) 1983. *cochlear prostheses.* An international symposium. Annals of the New York academy of sciences. Vol. 405

Parkins C.W. and Colombo J. 1987. Auditory-nerve single single-neuron thresholds to electrical stimulation from scala tympani electrodes. Hearing Res. 31, 267-286

Petrovsky J.S. 1978. Control of the recruitment and firing frequencies of motor units in electrically stimulated muscles in cat. Med. and Biol. Eng. and Comput. 16, 302-308

Plant R.E. 1978. The effects of calcium^{++} on bursting neurons. A modelling study. Biophys. J. 21, 217-238

Plonsey R. 1974. The active fiber in a volume conductor. IEEE-Trans. Biomed. Eng. BME-21, 371-381

Quick D.C., Kennedy W.R. and Donaldson L. 1979. Dimensions of myelinated nerve fibers near the motor and sensory terminals in cat tenuissimus muscles. Neuroscience 4, 1089-1096

Ranck J.B. jr. 1975. Which elements are excited in electrical stimulation of mammalian central nervous system: A review. Brain Res. 98, 417-440

Rattay F. 1986a. High frequency electrostimulation of excitable cells. J. theor. Biol. 123, 45-54

Rattay F. 1986b. Analysis of models for external stimulation of axons. IEEE-Trans. Biomed. Eng. BME-33, 974-977

Rattay F. 1987a. Modelling and simulation of electrically stimulated nerve and muscle fibers: A review. Math. and Comput. in Simulation. 29, 357-366

Rattay F. 1987b. Ways to approximate current-distance relations for electrically stimulated fibers. J. theor. Biol. 125, 339-349

Rattay F. 1988. Modeling the excitation of fibers under surface electrodes. IEEE-Trans. Biomed. Eng. BME-35, 199-202

Rattay F. 1989. Analysis of models for extracellular fiber stimulation. IEEE-Trans. Biomed. Eng. BME-36, 676-682

Rattay F. und Mayr W. 1987. Eine quantitative Abschätzung elektrisch aktivierter Fiberpopulationen am Beispiel der Karussellstimulation. Biomedizinische Technik 32, 184-190

Rattay F. and Motz H. 1987. Simulation of a multichannel nerve array to pulse shapes produced by single channel electrostimulation. Perception 16, 769-776

Reilly J.P. 1989. Peripheral nerve stimulation by induced electric currents: exposure to time-varying magnetic fields. Med. Bio. Eng. & Comp. 27, 101-110

Revenko S.V., Timin Y.E. and Khodorov B.I. 1973. Special features of the conduction of nerve impulses from the myelinized part of the axon into the non-myelinated terminal. Biophysics 18, 1140-1145

Ritchie J.M. and Rogart R.B. 1977. The binding of saxitoxin and tetrodotoxin to excitable tissue. Rev. Physiol. Biochem. Pharmacol. 79, 1-50

Roberts W.J. and Smith D.O. 1973. Analysis of threshold currents during microstimulation of fibres in the spinal cord. Acta physiol. scand. 89, 384-394

Röper J. and Schwarz J.R. 1989. Heterogeneous distribution of fast and slow potassium channels in myelinated rat nerve fibers. J. Physiol. 416, 93-110

Rose J.E., Brugge J.F., Anderson D.J. and Hind J.E. 1967. Phase-locked response to low-frequency tones in single auditory nerve fibers of the squirrel monkey. J. Neurophysiol. 30, 769-793

Roth B.J. and Basser P.J. 1990. A model of the stimulation of a nerve fiber by electromagnetic induction. IEEE-Trans. Biomed. Eng. BME-37, 588-596

Roth B.J. and Wikswo J.P. jr. 1985. The electrical potential and the magnetic field of an axon in a nerve bundle. Math. Biosci. 76, 37-57

Rushton W.A.H. 1951. A theory of the effects of fiber size in medullated nerve. J. Physiol. 115, 101-122

Rutherford K.M., Riso R. and Neuman M.R. 1987. A finger mounted force sensor for use with an FNS hand grasp system. IEEE 9-th ann. conf. Eng. Med. Biol. Soc. 1757-1758

Rutherford W. 1886. A new theory of hearing. J. Anat. Physiol. 21, 166-168

Ryall R.W. 1979. *Mechanism of drug action on the nervous system.* Cambridge: University Press

Rydevik B., Lundborg G. and Bagge U. 1981. Effects of graded compression on intraneural blood flow – An in vivo study on rabbit tibial nerve. J. Hand Surg. 6, 3-12

Scott A.C. 1977. *Neurophysics.* Interscience, Wiley, New York

Scriven D.R.L. 1981. Modelling repetitive firing and bursting in a small unmyelinated nerve fiber. Biophys. J. 35, 715-730

Schroeder M.R. 1975. Models of hearing. Proc. IEEE 63, 1332-1350

Schwarz J.R. 1986. The effect of temperature on Na currents in rat myelinated nerve fibres. Pflügers Arch. 406, 397-404

Schwarz J.R. and Eikhof G. 1987. Na currents and action potentials in rat myelinated nerve fibres at 20 and 37^oC. Pflügers Arch. 409, 569-577

Shaw E.A.G. 1974. Transformation of sound pressure level from the free field to the eardrum in the horizontal plane. J. Acoust. Soc. Amer. 65, 1848-1861

Sherratt R.M., Bostock H. and Sears T.A. 1980. Effects of 4-aminopyridine on normal and demyelinated nerve fibres. Nature 283, 570-572

Sigworth F.J. and Neher E. 1980. Single Na^+channel currents observed in cultured rat muscle cells. Nature 287, 447-449

Simmons F.B. 1966. Electrical stimulation of the auditory nerve in man. Arch. Otolaryngol. 84, 22-76

Singer R.S. and Nicolson G.L. 1972. The fluid mosaic model of the structure of cell membrane. Science 175, 720-731

Smith B., Peckham P.H., Keith M.W. and Roscoe D.D. 1987. An externally powered, multichannel, implantable stimulator for versatile control of paralyzed muscle. IEEE Trans. Biomed. Eng. BME-34, 499-508

Sneddown I.N. 1966. *Mixed boundary value problems in potential theory.* North Holland, Amsterdam

Sokolich W.G., Hamernik R.P., Zwislocki J.J and Schmiedt R.A. 1976. Inferred response polarities of cochlear hair cells. J. Acoust. Soc. Am. 59, 963-974

Solomonow M. 1984. External control of the neuromuscular system. IEEE-Trans. Biomed. Eng. BME-31, 752-763

Solomonow M., Baratta R., Zhou B.H., Shoji H. and D'Ambrosia R.D. 1987. The EMG-force model of electrically stimulate muscle: dependece on control strategy and predominant fiber composition. IEEE-Trans. Biomed. Eng. BME-34, 692-703

Spoendelin H. 1969. Innervation patterns in the organ of Corti of the cat. Acta Otolaryngol. 67, 239-354

Steel K.P. 1983. The tectorial membrane of mammals. Hearing Res. 9, 327-359

Sunderland S. 1978. *Nerves and Nerve Injuries.* Churchill Livingstone, New York.

Sweeney J.D. and Mortimer J.T. 1986. An asymmetric two electrode cuff for generation of unidirectionally propagated action potentials. IEEE-Trans. Biomed. Eng. BME-33 541-549

Sweeney J.D., Ksienski D.A. and Mortimer J.T. 1990. A nerve cuff technique for selective excitation of peripheral nerve trunk regions. IEEE-Trans. Biomed. Eng. BME-37, 706-715

Sweeney J.D., Mortimer J.T. and Durand D. 1987. Modeling of mammalian myelinated nerve for functional neuromuscular electrostimulation. IEEE 9-th ann. conf. Eng. Med. Biol. Soc. Boston. 1577-1578

Tanner J. 1962. Reversible blocking of nerve conduction by alternating current excitation. Nature 195, 712-713

Tasaki I. 1954. Nerve impulses in individual auditory nerve fibers of guinea pig. J. Neurophysiol. 17, 97-122

Tasaki I. 1955. New measurements of the capacity and the resistance of the myelin sheath and the nodal membrane of the isolated frog nerve fiber. Am. J. Physiol. 181, 639-650

Tay G., Chilbert M., Battocletti J., Sances A., Swiontek T. and Kurakami C. 1989. Measurements of magnetically induced current density in saline and in vivo. Proc. 11th Ann. Conf. IEEE-EMBS 1167-1168

Taylor R.E. 1963. Cable theory. In Physical techniques in biological research. Ed. Nastuk W.L. Academic Press, New York. Vol 6, 219-262

Ungar I.J., Mortimer J.T. and Sweeney J.D. 1986. Generation of unidirectionally propagating action potentials using a monopolar electrode cuff. Ann. Biomed. Eng. 14, 437-450

Van den Honert C. and Mortimer J.T. 1979. Generation of unidirectionally propagated action potentials in peripheral nerve by brief stimuli. Science 206, 1311-1312

Van den Honert C. and Mortimer J.T. 1981. A technique for collision block of peripheral nerve: Single stimulus analysis. IEEE Trans. Biomed. Eng. BME 28, 373-378

Van den Honert C. and Mortimer J.T. 1981. A technique for collision block of peripheral nerve: Frequency dependence. IEEE Trans. Biomed. Eng. BME 28, 379-382

Van den Honert C. and Stypulkowski P.H. 1987a. Single fiber mapping of spatial excitation patterns in the electrically stimulated auditory nerve. Hearing Res. 29, 195-206

Van den Honert C. and Stypulkowski P.H. 1987b. Temporal response patterns of single auditory nerve fibers elicited by periodic electrical stimuli. Hearing Res. 29, 207-222

Veltink P.H., Alste J.A. van and Boom H.B.K. 1988. Simulation of intrafascicular and extraneural nerve stimulation. IEEE-Trans. Biomed. Eng. BME-35, 69-75

Volta A. 1800. Letter to Sir Joseph Banks, March 20, 1800. On electricity excited by the mere contact of conducting substances of different kinds. Phil. Trans. Roy. Soc. 90, 403-431

Vossius G. 1983. Functional muscle stimulation with surface electrodes. 36th ACEMB, Columbus Ohio.

Vossius G. 1986. Comparison of FES of the upper extremities by means of surface and implanted electrodes. Proc. 2nd Vienna int. workshop on functional electrostimulation. 117-119

Vossius G., Müschen U. and Holländer H.-J. 1987. Multichannel stimulation of the lower extremities with surface electrodes. 9th int. symposium on advaces in external control of human extremities. Dubrovnik, 193-203

Walsh E.G. 1964. *Physiology of the nervous system.* Longman, Harlow

Wedensky N. 1884. Wie rasch ermüdet der Nerv? Zbl. med. Wiss. 65-68

Weiss G. 1901. Sur la possibilite de rendre comparable entre eux les appareils servant a l'excitation electrique. Arch. ital. Biol. 35, 413-446

Wever E.G. 1954. *Theory of Hearing*, Dover, New York

Wever E.G. and Lawrence M. 1954. *Physiological Acoustics.* Princeton Univ. press. Princeton N.J.

Wiley J.D. and Webster J.G. 1982. Analysis and control of the current distribution under circular dispersive electrodes. IEEE-Trans. Biomed. Eng. BME-29, 381-385

Woo N. and Campbell B. 1964. Asynchronous firing and block of peripheral nerve conduction by 20 kc alternating current. Bull. L.A. Neurol. Soc. 29, 87-94

Woosley J.K., Roth B.J. and Wikswo J.P. jr. 1985. The magnetic field of a single axon: a volume conductor model. Math. Biosci. 76, 1-36

Young J.Z. 1936. Structure of nerve fibres and synapses in some invertebrates. Cold Spring Harbor Symp. Quant. Biol. 4, 1-6

Zhou B.H., Baratta R. and Solomonow M. 1987. Manipulation of muscle force with various firing rate and recruitment control strategies. IEEE-Trans. Biomed. Eng. BME-34, 128-139

Zienkiewicz O.C. 1971. *The finite element method in engineering science.* McGraw-Hill, London

AUTHOR INDEX

SUBJECT INDEX*

---

* Bold faced numbers refer to detailed explanations.

Druck: Novographic, Ing. Wolfgang Schmid, A-1230 Wien.

GPSR Compliance

The European Union's (EU) General Product Safety Regulation (GPSR) is a set of rules that requires consumer products to be safe and our obligations to ensure this.

If you have any concerns about our products, you can contact us on ProductSafety@springernature.com

In case Publisher is established outside the EU, the EU authorized representative is:

Springer Nature Customer Service Center GmbH
Europaplatz 3
69115 Heidelberg, Germany

Batch number: 10577876

Printed by Printforce, the Netherlands